Smart Textiles Production

Inga Gehrke, Vadim Tenner, Volker Lutz,
David Schmelzeisen and Thomas Gries

Smart Textiles Production
Overview of Materials, Sensor and Production Technologies for Industrial Smart Textiles

MDPI • Basel • Beijing • Wuhan • Barcelona • Belgrade

MDPI

AUTHORS

Inga Gehrke, Vadim Tenner, Volker Lutz, David Schmelzeisen and Thomas Gries

Institute of Textile Technology

RWTH Aachen University

Germany

EDITORIAL OFFICE

MDPI

St. Alban-Anlage 66

Basel, Switzerland

For citation purposes, cite as indicated below:

Gehrke, I.; Tenner, V.; Lutz, V.; Schmelzeisen, D.; Gries, T. *Smart Textiles Production. Overview of Materials, Sensor and Production Technologies for Industrial Smart Textiles*; MDPI: Basel, Switzerland, 2019.

FIRST EDITION 2019

ISBN 978-3-03897-497-0 (Hbk)
ISBN 978-3-03897-498-7 (PDF)

doi:10.3390/books978-3-03897-498-7

Cover image courtesy of Inga Gehrke, Vadim Tenner, Volker Lutz, David Schmelzeisen and Thomas Gries.

Contents

PART I
by Volker Lutz and Inga Gehrke

Introduction to Smart Textiles: Applications and Markets

The term "Smart Textiles" has now reached the general public and has massively increased the demand for new, functional textile products. The market research company IDTechEx predicts a market of approximately €2.8 billion for 2026 with an average annual growth rate of 34% [1].

Smart Textiles are textiles with an extended range of functions. An essential goal of the extended functional scope is the interaction of the textile with the environment, which also includes the human user. The European Committee for Standardization (CEN) defines Smart Textiles in the technical report (TR) 16298:2011 more specifically as intelligent systems consisting of textile and non-textile components that actively interact with their environment, a user or an object (Figure 1). Data is recorded and processed via sensors and a defined reaction is generated via actuators or an information display on an additional device [2].

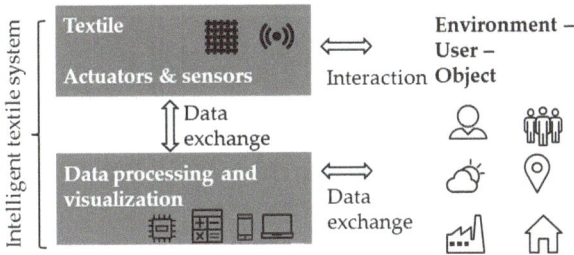

Figure 1. Schematic representation of Smart Textiles as an intelligent textile system, according to Reference [2].

Especially in combination with digital networked services, Smart Textiles promise support in almost all situations (Figure 2). Above all, the possible applications in sports, health, home and living, mobility or building open up completely new markets and business models for both consumer and technical products.

(a)   (b)

Figure 2. Smart Textile prototypes with the adaption of electronics for lighting (**a**) and interaction/sensing applications (**b**).

In addition to other products based on flexible or portable electronics, textile-based electronics promise an established user acceptance, since textiles are the most common material in the human environment, whether close to the body or directly surrounding it.

Like comparable technologies, Smart Textiles are subject to an initial euphoria (Gartner Hype Cycle [3]) followed by a rapid disillusionment of the market due to a lack of marketability. A major challenge is the lack of production technologies that can enable scalability from prototypes to marketable Smart Textiles. Moreover, when selecting the technologies used, not only the functionality but also the entire life cycle must be taken into account. In addition to usage, the requirements of product development from design to production must also be taken into account.

To date, there are only functional constructions for demonstrative purposes ("demonstrators") for most Smart Textiles. These receive a lot of attention. Unfortunately, however, such products are not available on the market in high volumes at short notice. Table 1 gives an overview of the most important application areas and product categories for Smart Textiles.

Demonstrators from these categories are often the result of hours of manual work. An economic transfer of production fails at the interfaces in the manufacturing steps in the various technical sub-areas addressed by Smart Textiles. Textile technology, electrical engineering and information technology have so far required different approaches, and there is a lack of common standards. A combination of the interdisciplinary competences, and the illustration of adequate division of partial steps, are necessary.

Table 1. Application fields and common product categories of Smart Textiles.

Application Fields	Product Categories	
Medicine and Health	• Monitor vital signs (blood pressure, heart rate, electrocardiography (ECG), electroencephalogram (EEG), blood sugar) • Wound healing monitoring	• Patches (drug delivery) • Motion analysis
Sport and Fitness, Wellness	• Monitor activity (steps, heart rate) • Stress monitoring	• Muscle stimulation • Sleep tracking
Industry and Military	• Protective equipment (PPE) • Ergonomics improvement • Counterfeit protection	• Monitor attention • Exoskeletons
Home and Architecture	• Integrated displays and control • Structural health monitoring for buildings	• Activity tracking (movement, falls)
Fashion, Lifestyle, Others	• Integrated displays and outdoor control	• Visual and haptic effects

The lack of market breakthrough for Smart Textiles is thought to be a result of the following technology-related aspects:

- Depending on the usage requirements, Smart Textiles have to survive mechanical, chemical and thermal treatments over their life cycle, e.g., washing, ironing, tumbling, stretching, abrasion, etc.
- In most cases, Smart Textiles need to be powered by portable energy sources such as batteries or energy harvesting technologies. Flexible batteries and energy harvesting technologies (e.g., photovoltaics, piezoelectrics, etc.) suffer from low energy output, low flexibility and insufficient human skin compatibility.
- Most industrial production technologies are not compatible with Smart Textile manufacturing. The upgrading of the existing production processes from the lab to the industrial scale is considered not to be economical.

This book focuses on production technologies for integrating the components of an intelligent textile system into a Smart Textile product. Both aspects are treated separately and are covered by strong international research and development activities.

The current industrial landscape strongly resembles an individual prototype production for the numerous product diversifications. As a result, an enormous number of manual work steps is required, which massively increases production costs in high-wage countries and thus results in low market penetration of the otherwise innovative products.

In this book, knowledge of previous Smart Textiles is summarized in order to support future work in the development and implementation of Smart Textiles. The aim is to describe Smart Textiles in a structured way with regard to materials that are available for product developments (Part II), functionality with focus on textile-based sensors (Part III), production technologies for integrating these functionalities into products (Part IV) and product concepts along the example of touchpads (Part V). It will help designers to understand the possible methods of Smart Textile production, so that they are enabled to design their products for scalable production. Moreover, it will assist textile and electronics manufacturers to decide which production technologies are suitable for meeting certain product requirements, thus contributing to reducing market entry thresholds.

PART II

by Inga Gehrke, Vadim Tenner and David Schmelzeisen

Electrically Conductive Fibers for Textiles

One of the most important properties of materials for "Smart Textiles" is their electrical conductivity. This chapter gives an overview of conductive polymers and silver coatings, as these are representative of the raw materials that feed the production technologies presented in Part IV. The overview is based on References [4–6] unless stated otherwise.

Conductive fibers can be intrinsically conductive, depending on the material, or alternatively extrinsic conductivity can be achieved by additional processing steps, as shown in Figures 3 and 4.

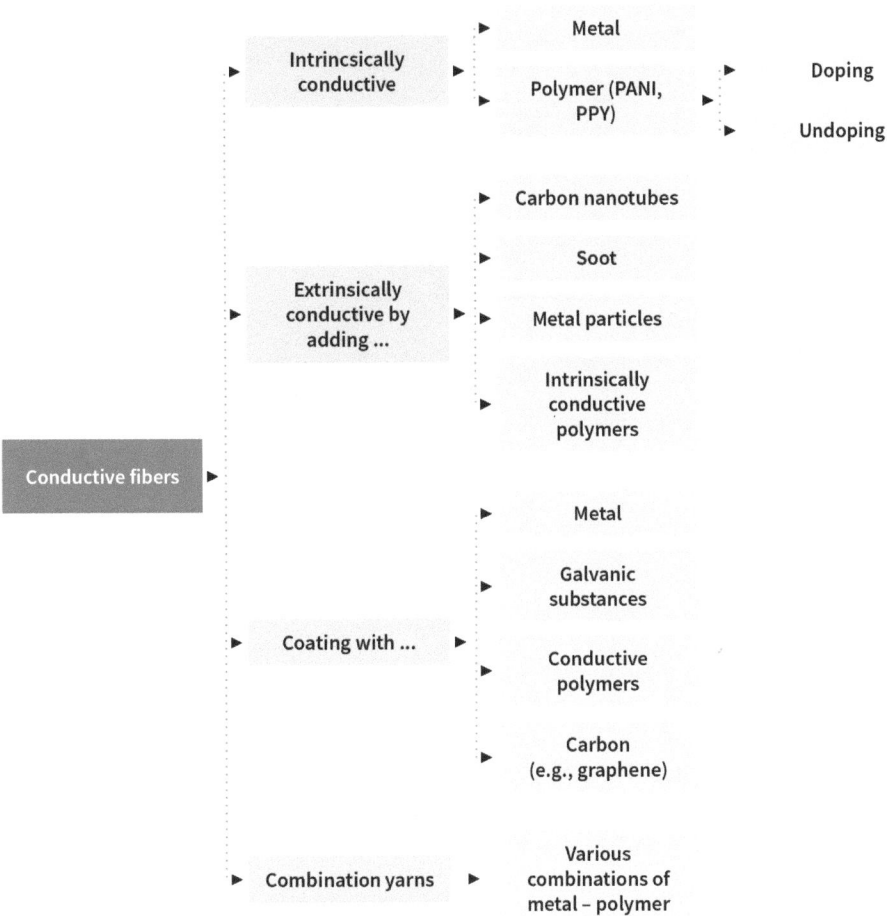

Figure 3. An overview of conductive fibers. PANI: Polyaniline; PPY: Polypyrrole.

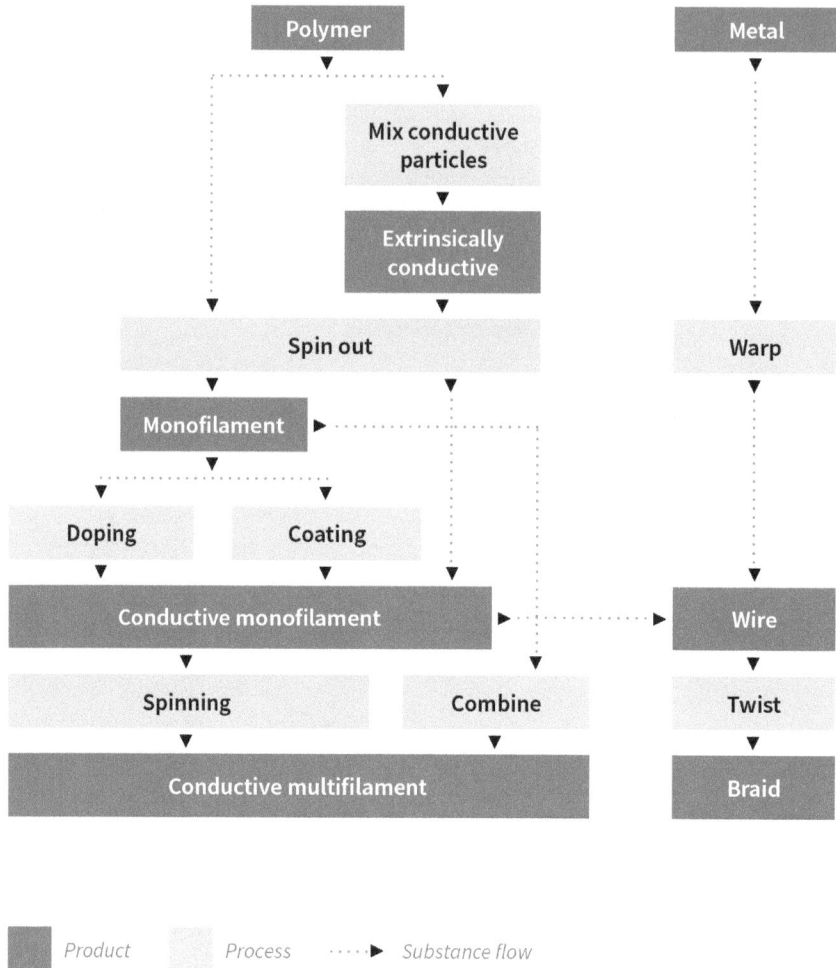

Figure 4. Schematic of the production processes for the manufacture of electrically conductive yarns.

While the primary research subjects for electrically conductive fibers are carbon nanotubes and graphene, e.g., Reference [7], industrial applications usually use conductive polymers or metal-coated yarns, which are described in more detail below.

2.1. Conductive Polymers

Plastics are usually lightweight, durable, easy to form and process, and inexpensive to manufacture. Due to their chemical structure, polymers are perfect insulators against electricity, i.e., exactly the opposite of metals. Under these conditions, it should be paradoxical to assume that it is the plastics that conduct the current. However, the U.S. researchers Alan Heeger and Alan MacDiarmid, together with the Japanese researcher Hideki Shirakawa, who were collectively awarded the Nobel Prize in Chemistry in 2000, discovered how to construct and treat polymers so that they become electrically conductive [8].

In order for electrons to move freely in plastics, rather than being coupled to atomic nuclei as is usually the case, they must alternately form single and double bonds between carbon atoms (conjugated double bonds). In polyacetylene, which is produced from the gas acetylene, these structural elements are perfectly combined to form a "conjugated" chain. Polyacetylene had long been known as a black powder when, in the early 1970s, Shirakawa and a colleague discovered how to synthesize polyacetylene in a new way and obtain black films that could be peeled from the inner wall of the reaction vessel. Additionally, they oxidized (or doped) polyacetylene with chlorine, bromine and iodine, and were thus able to increase the conductivity to 103 S cm^{-1}, which is in the range of semiconductors and metals. A disadvantage of polyacetylene is its air sensitivity; the initially very good conductivity quickly decreases due to reactions with oxygen. One way to solve this problem is to use doped polyacetylene as a component of specially manufactured polymer blends with thermoplastics, e.g., as an antistatic transparent film. In this form, the polyacetylene is better protected against aging.

Today, conductive plastics are used as antistatic films, electromagnetic shielding in electronic circuits, screen protectors, in through-plated circuit boards in the electronics industry, and in corrosion protection.

2.1.1. Polyaniline (PANI)

Considered a "metallic" plastic, polyaniline (PANI) is highly crystalline, largely chemically inert, and electrically conductive (it contains many free electrons). U.S. Army stealth jets are invisible to radar because, among other reasons, they are coated with a conductive PANI layer that completely absorbs the microwaves emitted by radar instead of reflecting them. Another application is color displays with minimal power consumption, which shine up to 100 times brighter than conventional color screens.

Due to its internal structure, PANI also appears to be very suitable for applications in nanotechnology. It can be divided into so-called primary particles 7–15 nm in size (this refers to the smallest units that possess all the properties of the plastic). In order to be able to process it at all, the PANI produced as a powder must be dissolved in water. Depending on the application, dispersions with a PANI content of up to 2% by weight are used.

In the medium, the PANI automatically forms a web structure, similar to a spiderweb. A total of 2% of PANI is required for dispersions for corrosion protection, 1% for the production of light-emitting diodes, and only 0.1% for the production of solderable surfaces on circuit boards. The particle sizes also vary, and range from 10–30 nm for electronic components to about 70 nm for anti-rust paints.

2.1.2. Poly(ethylenedioxythiophene) (PEDOT)

Major progress towards the industrial application of conductive polymers was achieved with the development of poly(ethylenedioxythiophene) (PEDOT or PEDT) by Bayer in the early 1990s. Thanks to its chemical structure, it is the most stable of all known conductive polymers and is used as a thin antistatic layer in photographic films made by the Bayer subsidiary Agfa-Gevaert N.V. The annual production of many hundreds of thousands of square meters of these ultra-thin layers requires only a few thousand kilograms of polymer.

2.1.3. Poly(3,4-ethylenedioxythiophene):poly(styrenesulfonate)

Another example of electrically conductive polymers with a wide range of applications, e.g., in optoelectronic devices [9], is poly(3,4-ethylenedioxythiophene) doped with poly(styrenesulfonate) anions, also called PEDOT:PSS.

PEDOT:PSS is a blend of cationic polythiopene derivative, doped with a polyanion. After doping with suitable solutions and the associated significant increase in electrical conductivity, PEDOT:PSS can be used as a transparent electrode and thus as an alternative to the frequently used indium tin oxide (ITO). With a conductivity of up to 4600 S cm^{-1}, it can also be used as cathode material in capacitors [10].

Due to its high electrical conductivity and good oxidation resistance, PEDOT:PSS can be used to coat textile substrates for applications such as electrodes for electrocardiographs, electrical and chemical transistors, electrodes for organic solar cells and organic light-emitting diodes (OLEDs) [11].

For example, the production of electrocardiography (ECG) electrodes using PEDOT:PSS is explained based on a paper by Pani et al. [12]. A solution of a PEDOT:PSS dispersion and a second donor is immersed in cotton or polyester fabric for 48 h. The textile is then pressed and heat-treated to remove the dispersion and evaporate the second donor or water [12].

Compared to the conventional Ag/AgCl electrodes used for ECG, PEDOT:PSS electrodes have the advantage that they function both when dry and wet. In principle, their conductivity is comparable to or better than that of the conventional electrodes. The disadvantage of PEDOT:PSS electrodes is that they have a higher contact impedance due to the material and their irregular surface. [12].

PEDOT:PSS can also be used to coat yarns and apply conductor paths to textile substrates using conventional methods such as sewing and embroidery. In 2017, Ryan et al. produced a PEDOT:PSS-coated silk yarn up to 40 m in length with an Young's modulus of 2 GPa and an electrical conductivity of 14 S cm^{-1}. Washing and drying cycles were possible, even if limited, without loss of conductivity. Ethylene glycol (EG), dimethyl sulfoxide (DMSO) and methanol 99% (MeOH) were used as second donors [11].

In 2015, Åkerfeldt et al. showed a process in which a conductive PEDOT:PSS solution is printed onto a textile by screen-printing by adding a binder to create a paste [13]. The solution is an aqueous dispersion of self-crosslinking acrylic with a solids content of 47.5 w%. For the textile coating, a PEDOT:PSS solution is mixed with a commercial binder and a polyurethane-based thickening agent. Ethylene glucol is used as the second donor. Compared to conventional polyurethane-based pastes, this paste offers the advantage of containing no metallic particles. Silver particles have a proven negative effect on their environment if they escape from the textile during washing or wearing [13].

2.2. Silver Coating

Silver-coated yarns are also used as electrical conductors. These include, for example, silver-coated polyamide multifilaments. Such yarns can be very easily processed in all textile processes, e.g., warp- and weft-knitting, weaving and embroidery. The high conductivity of silver makes it suitable for energy and data transmission through textiles (see Section 4.2).

In contrast to antibacterial applications, silver coatings are sensitive to washing processes and other mechanical stresses when used as electrical conductors. Due to their stable conductivity and good processability, industrial applications usually use fibers coated with silver or copper, such as the product brands Elitex, Shieldex and SEFAR [4].

PART III

by Inga Gehrke and Patrycja Bosowski-Schoenberg;
design and illustrations of catalog by Jan Serode

Classification of Textile-Based Sensors for Developing Smart Textiles

While numerous textile-based sensors have been developed, ranging from sensing fibers to coatings and three-dimensional structures, transparency regarding their specific properties and usage is missing. Bosowski et al. have suggested a structure for a classified catalog as a knowledge basis to support the smart textile product development process [14]. This chapter develops the classification further and implements it as a catalog to be used by practitioners from research and industry when developing and designing textiles with sensing capabilities. The appendix holds the full catalog.

3.1. Motivation: Need for Classified Knowledge on Textile-Based Sensors

In addition to the definition of Smart Textiles as "intelligent textile system" given by the CEN/TR 16298:2011 standard (see Part I), Smart Textiles can be classified according to the degree of textile integration. This describes the extent to which electronic components are covered by textiles (Figure 5).

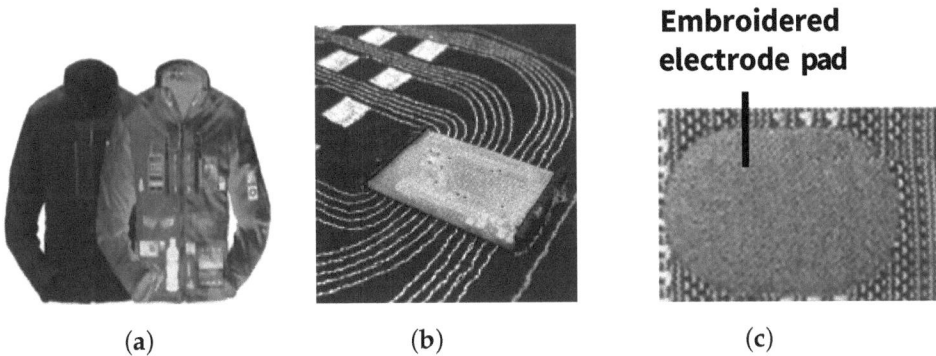

Embroidered electrode pad

(a)   (b)   (c)

Figure 5. Classification of Smart Textiles according to the degree of textile integration [15]. (a) Textile-adapted; (b) Textile-integrated; (c) Textile-based.

Textile-adapted: the textile does not cover an electronic function (0%), but electronic components can be attached to the textile, e.g., a pocket for an MP3 player.

Textile-integrated: the textile covers between 0 and 100% of the electronic function, creating an interface between the textile and the electronics. For example, flexible circuit boards can be integrated into textiles in this manner.

Textile-based: electronic function is 100% covered by the textile. When considering an intelligent textile material, this can involve the realization of conductor paths and sensors made of conductive yarns.

Textile-based sensor technology is a critical component for the functionalization of textiles, as it offers the promising possibility of integration into existing textile structures from everyday life such as clothing or interior design. Due to the high degree of textile integration, the electronic function can "disappear" into the textile and thus be worn discreetly and inconspicuously. At the same time, textile-based implementations allow a pleasant feel for the user [16]. Despite these advantages, there are hardly any Smart Textiles available on the mass market. Technological barriers need to be addressed, such as robust integration and contacting technologies that can withstand the usage requirements of clothing (washing, tumble drying, ironing) while conforming with non-toxicity certifications, and the improvement and miniaturization of power management and storage devices [17]. Part VI describes challenges and solution approaches for scalable production processes in more detail. Even where technological progress has been made, lack of transparency about available components, their application possibilities and their degree of maturity complicates the product development process and market launch, especially since developers with an electronics background typically have limited experience with textile-based components and vice versa [17]. The search for a textile sensor and its design for a special application, as already attempted by many research projects (cf. [15,18–24]) has so far involved many examinations of thread combinations and materials. This is a lengthy and costly process. This has already generated knowledge about textile sensor technology, which requires appropriate classification and structure. This is implemented here with a design catalog that implements design principles according to Reference [14] and is expanded according to the current state of research. It is intended to serve developers of Smart Textiles as an information basis for the selection of textile-based sensor modules and thus contribute to the faster and more successful market launch of Smart Textiles.

3.2. Textile-Based Sensors within the Context of Smart Textiles

3.2.1. Areas of Application for Textile-Based Sensors

Textile-based sensor technology offers a wide range of possible applications by considerably expanding the range of functions of ordinary textiles by combining them with sensors. In addition to applications close to the body in clothing, where sensors are usually used to monitor vital functions and movements (cf. [15,18–24]), textile-based sensors are also used to monitor their carrier material in the construction industry. Examples are load investigations of structures or the monitoring of slope fixings in dams and dikes [14,15].

Figure 6 shows the fields of application based on the definition of basic terms for technical textiles according to Gries et al. [25].

APPLICATION AREA	EXAMPLES
Agrotech	Gardening and landscaping, agriculture and forestry, animal husbandry
Buildtech	Engineering and industrial construction, interior fittings, earthworks, waterworks and traffic route construction
Clothtech	Clothing, shoes
Geotech	Civil engineering, road construction, dam construction, landfill construction and mining
Hometech	Furniture, upholstery, interior equipment, floor and wall coverings
Indutech	Filtration, cleaning, mechanical engineering, chemical and electrical industry
Packtech	Packaging, insulation, covering
Protech	Personal, building and property protection
Medtech	Medicine, hygiene, work clothing, implants
Mobiltech	Vehicles of all kinds, ships, aviation and space travel
Ecotech	Environmental protection, recycling, disposal
Sporttech	Sport and leisure, functional sportswear

Figure 6. Fields of application of textile engineering according to Gries et al. [25].

3.2.2. Definition of a Textile-Based Sensor

Textile-based sensors always consist of a textile material and are defined by their textile structure. The sensor can be incorporated into a textile substrate—the so-called supporting textile—which is not itself a part of the textile sensor in this case [14].

A textile sensor can be defined over four levels of the textile structure as well as according to its manufacturing process and its material, which must be specifically selected depending on the type of textile (Figure 7).

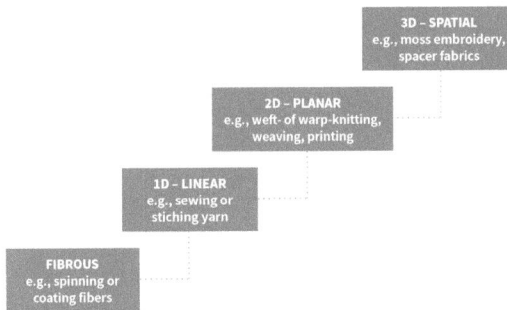

3D – SPATIAL
e.g., moss embroidery, spacer fabrics

2D – PLANAR
e.g., weft- of warp-knitting, weaving, printing

1D – LINEAR
e.g., sewing or stiching yarn

FIBROUS
e.g., spinning or coating fibers

Figure 7. Levels of textile structure and related manufacturing processes.

Depending on the application, sensors from different manufacturing levels with suitable materials are required. At the fiber level, fibers that conduct light or electric current serve as the basis for sensor functions. An overview of production processes for conductive fibers and their properties can be found in Part II and elsewhere, e.g., Reference [4]. In the one-dimensional plane, thread-shaped sensors should be mentioned; these are inserted into supporting textiles in the form of a yarn or thread by sewing or embroidering in a linear manner. The next level is flat textiles, e.g., warp- and weft-knitted or woven fabrics. Planar structures also result when conductive pastes (e.g., silver pastes) are printed on the supporting textile (see Chapter 6.3.2 for a detailed description of printing technologies). Three-dimensional (spatial) structures are created when sensors are inserted into spacing weaves or warp knits or are inserted via 3D embroidery (moss embroidery) [15].

3.3. Classification of Textile-Based Sensors

3.3.1. Objective of Cataloging Textile-Based Sensors

The catalog is intended to support developers and designers as an information database in the design of textile-based sensors by providing proposals for their methodical use in an application-friendly manner. This catalog was therefore designed in accordance with the Verein Deutscher Ingenieure (VDI, Association of German Engineers) Guideline VDI 2222, Part 2, "Preparation and Application of Design Catalogues", of February 1982 [26]. It illustrates the networking of the individual criteria for the correct design methodology procedure in the creation and application of a design catalog (Figure 8).

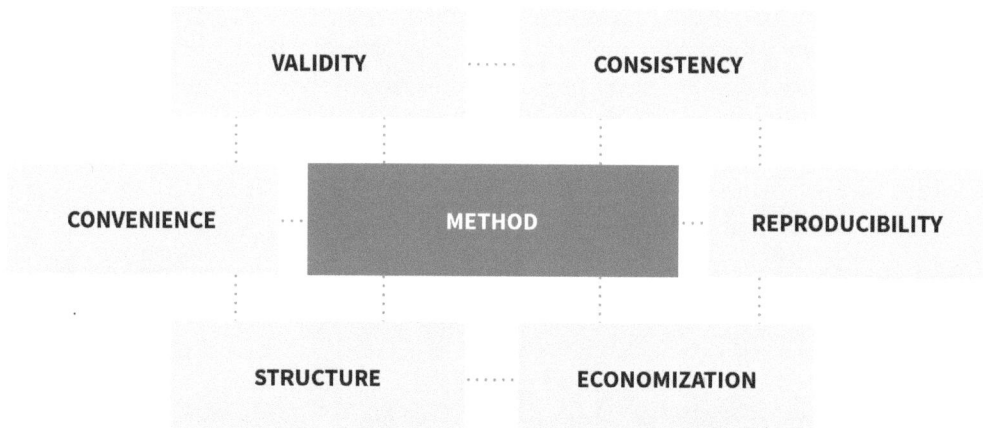

Figure 8. Approach for construction catalog design according to the Verein Deutscher Ingenieure (VDI, Association of German Engineers) [26].

Without claiming to be complete—design catalogs require constant updating—the tables present a combination of the current state of textile sensor technologies. They also offer designers the opportunity to implement reproducible design processes independently of their own knowledge and to achieve rationalization effects thanks to the efficient provision of information [26].

In this respect, a structuring of the catalog according to aspects of design methodology seems plausible, and allows direct and targeted access with the simultaneous provision of structured information [26].

This catalog is subject to the requirement of ensuring not only comfortable handling, quick access to information and consideration of construction methodological terms and procedures, but also completeness and validity for as many users as possible. Furthermore, its design is consistent. Finally, the catalog must be systemically consistent, however its details may be changed [26]. Each user of this catalog is encouraged to add to and extend it in good conscience, taking into account the abovementioned requirements.

3.3.2. Implications for Designing a User-Oriented Catalog

For solution catalogs, the degree of concretization of the solution description is a critical design parameter.

Deciding on a generally valid formulation of the solution implies that the designer is able to abstract the problem and implement it analogously to the description in the catalog. Only the initial phase of the design process is covered in the catalog, whereas the actual problem-solving process remains unaffected by the catalog, as its implementation requires too many concrete details [26]. Figure 9 illustrates the relationship between the four sections of the problem-solving process (1. Problem; 2. Model; 3. Adjustment; 4. Implementation). It has been shown that in the case of a general representation of the solution for the implementation of the problem in a problem-solving process, a reduction of the degree of concretization is necessary. This transformation is application-specific, and consequently the catalog cannot capture most of the process.

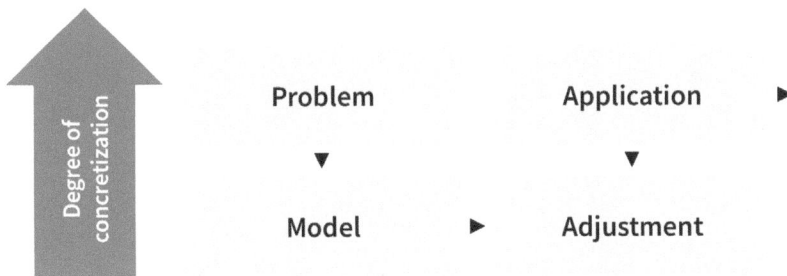

Figure 9. Problem-solving process with general formulation of the solution.

However, specific solutions can describe the entire construction process and thus be of assistance to the designer. Nevertheless, at the same time there is a danger that the more concrete the information becomes the more complex it will become, i.e., more information needs to be presented. This is expressed negatively in a large number of sub-functions, with details to be described [26].

In order for the catalog to support developers in all application areas of technical textiles, the solution descriptions are based on the principle of being formulated as abstractly as possible. At the same time, developers should be supported in assigning the abstract solution to a specific problem by subdividing the textile sensors into application areas. Moreover, classifying the maturity level makes it easier for the developer to evaluate the feasibility of a solution.

3.3.3. Structure and Classification Method

In practice, there are often multiple ways to access the catalog contents. A comparison of several different characteristics between problem and solution description requires a clear and equal arrangement of the characteristics. It is suitable to apply all the solutions on one axis and to compare all the characteristics without repetition on a second axis. A linear (one-dimensional) arrangement of the solutions is therefore recommended for the catalog [26].

Primary classification aspects for the catalog are the classes of functions, which result from the various application areas of textile sensor technology (cf. Figure 6). The first column group comprises the classification criteria, which are intended to give the user a quick overview of the relevant solutions (Figures 10 and 11). These forms of access characteristics are arranged one after the other according to their meaning.

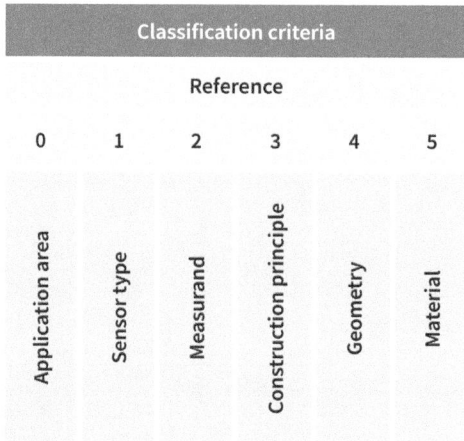

Figure 10. Classification criteria in the catalog.

The description of the sensor type (mechanical, chemical or thermal) follows the overriding criterion of the area of application. A specification of measuring principles, including an indication of the measured variables, enables a further narrowing of the solution. The final specification of the manufacturing principle, the textile geometry and the materials used makes the application field of the sensor more understandable.

ACCESS LEVEL 0 Application area	• Agrotech • Buildtech	• Clothtech • Geotech	• Hometech • Indutech	• Medtech • ...
FIRST LEVEL Sensor type	• Chemical • Mechanical	• Thermal		
SECOND LEVEL Measurand	• Electromagnetic light spectrum • Electric current		• Visual assessment	
THIRD LEVEL Construction principle	• Fiber • Thread	• Fleece • Weft knits	• Weave • Warp knits	• Scrim • ...
FOURTH LEVEL Geometry	• Punctiform • Linear	• Planar • Three-dimensional		
FIFTH LEVEL Material	• Cotton • Polymer	• Glass fiber • Carbon fiber	• Viscose • Cellulose	• Metals • ...

Figure 11. Structuring and contents of the classification criteria.

The solution area contains the essential information about the solutions (Figure 12). The procedural principle already covers the concise aspects of the function and the structure of the sensor. The solution sector is further illustrated by a schematic representation (schematic sketch).

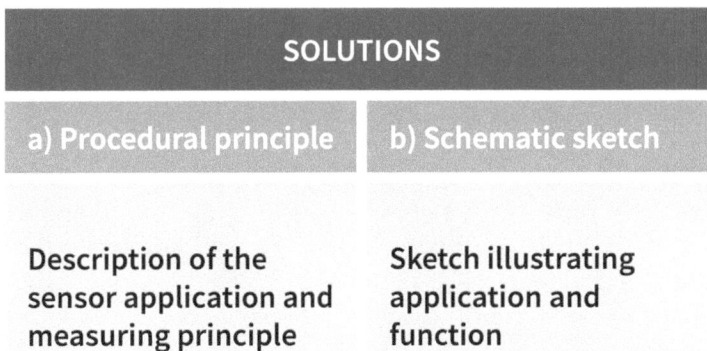

SOLUTIONS	
a) Procedural principle	b) Schematic sketch
Description of the sensor application and measuring principle	Sketch illustrating application and function

Figure 12. Solution area in the catalog.

The section on access features (Figure 13) supplements the information from the previous core area of the catalog by mentioning known application examples, conceivable variation possibilities and the advantages and disadvantages resulting from the use of the chosen sensor. Furthermore, this section includes characteristic properties and characteristic values of the textile, which can significantly narrow the selection of solutions. In particular, these access features are subject to constant optimization in the form of additions and extensions, which are carried out by the user on the basis of practical experience gained.

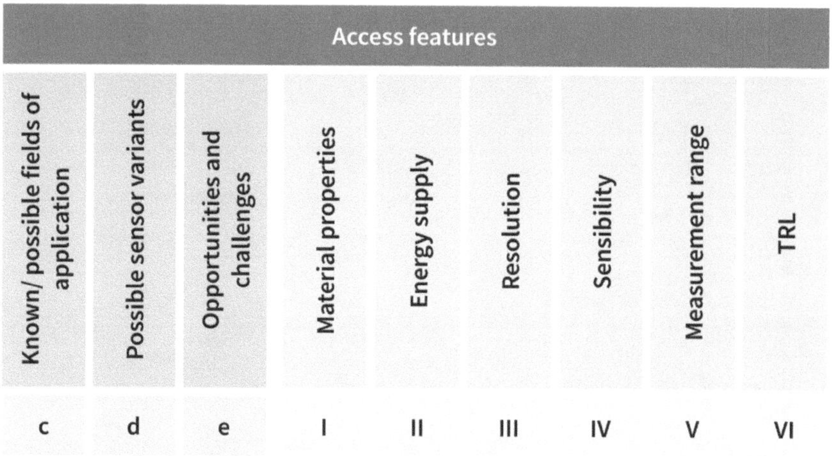

Figure 13. Access features in the catalog. TRL: Technology Readiness Level.

The degree of maturity of the technology, which is assessed via the "Technology Readiness Level" (TRL) [27], is of particular importance here. A classification is made according to the groups shown in Figure 14:

- Proof of concept/test in the laboratory environment (TRL $\leq$ 5)
- Demonstrator of the complete system in application environment (TRL = 6–8)
- Product used in application environment (TRL = 9)

This provides the user with an immediate overview of whether the textile-based sensor is ready for its application environment. For example, a TRL of 9 in "clothtech" applications indicates that operational stability for clothing has been shown, with detailed information given in catalog sections c and e (e.g., stability against washing shown, but not tumble drying). Most of the textile-based sensors described in the appendix are currently at a TRL of 6–8, as the required use in an application environment is often still hindered by the lack of operational stability (cf. Chapter 3.1).

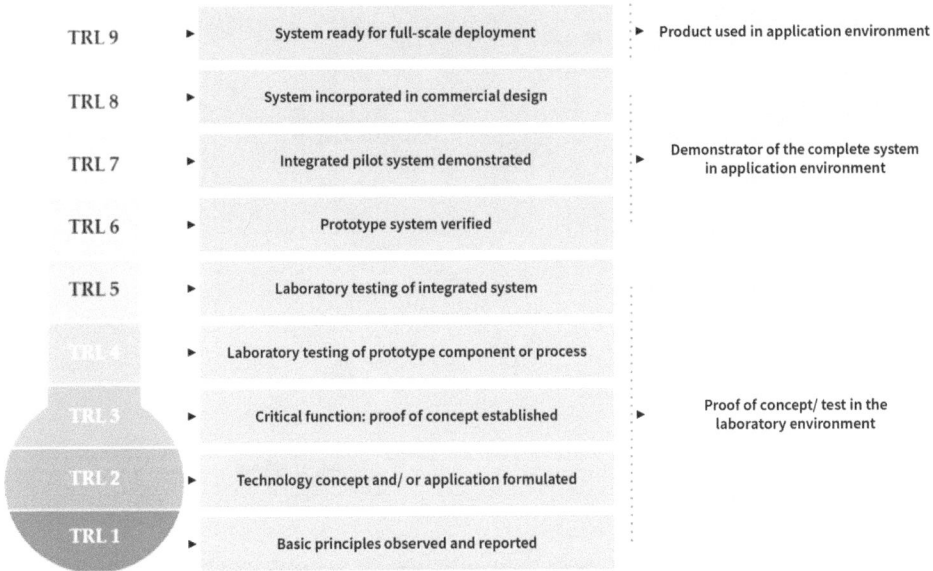

TRL 9	▸	System ready for full-scale deployment	┊ ▸ Product used in application environment
TRL 8	▸	System incorporated in commercial design	┊
TRL 7	▸	Integrated pilot system demonstrated	┊ ▸ Demonstrator of the complete system in application environment
TRL 6	▸	Prototype system verified	┊
TRL 5	▸	Laboratory testing of integrated system	┊
TRL 4	▸	Laboratory testing of prototype component or process	┊
TRL 3	▸	Critical function: proof of concept established	┊ ▸ Proof of concept/ test in the laboratory environment
TRL 2	▸	Technology concept and/ or application formulated	┊
TRL 1	▸	Basic principles observed and reported	┊

Figure 14. Technology Readiness Level according to Mankins [27] and reduced classification for this catalog.

3.3.4. Application Example: Developing a Health-Monitoring Evacuation Mattress Using the Proposed Classification

The structure and benefits of the catalog along the problem-solving process from Figure 9 are explained below using the example of a sensor developed in the "KostBar" research project [28,29].

Problem: In order to increase the safety of patients in hospitals and nursing homes, fall prevention and decubitus (pressure sore) prophylaxis in patient beds must be improved. These functions are to be realized via a sensor system integrated in evacuation mats that lie beneath the mattress in the patient's bed.

Model: Changes in the pressure distribution are to be measured over the mattress surface. From the suitable manipulation of the data obtained through the measured values, the approximate position of the patient on his or her bed can be determined. A change in position towards the edge of the bed should be used to warn against falls, while a warning due to a long-unchanged position serves as decubitus prophylaxis (Figure 15).

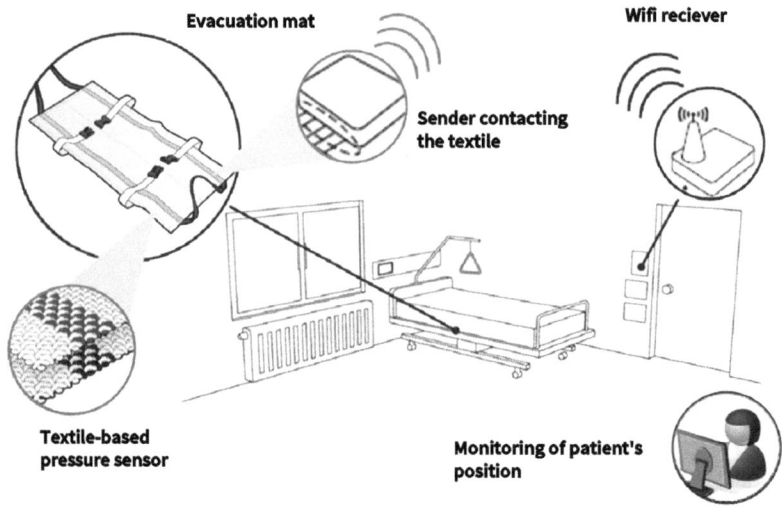

Figure 15. Model for the use of 3D tubular fabric for pressure measurement in the "KostBar" project [28].

Matching with catalog: Figure 16 shows the classification in the catalog in the "Medtech" area. It is also listed in the "Hometech" section to facilitate quick access. The descriptions of the solution area and the access characteristics are identical. TRL is rated at 6–8 as the demonstrator has been tested in an application environment but no permanent use has been reported. Thus, future users must know that operational stability testing will be required.

Implementation: A pressure sensor was attached to an evacuation mat in the form of 150-cm-long tubes with eight flat conductors and put into contact with an electronic unit (power supply and data transmission) to suit the problem. A capacitance measurement of the three-dimensional fabric sensor was recorded every 250 s. Any possible deviation greater than 2%, resulting from the simultaneous statistical processing of the data obtained, was transmitted to a receiver unit. The measurement results are evaluated and an alarm is triggered if a fall or decubitus is suspected [29].

This example shows that the structure of the catalog supports the problem-solving process in a targeted way: the classification according to application areas enables the user to find similar solutions to a problem. At the same time, the abstract description of the solution is not too restrictive: falls and decubitus prophylaxis are indeed the problems solved in the project. However, the tubular fabric is not limited to this usage, and users of the catalog should be enabled to evaluate its suitability for other problems. Access features such as technological maturity and application area are particularly helpful in this respect.

	3D TUBULAR FABRIC
1 | SENSOR TYPE	Mechanical
2 | MEASURAND	Electric current
3 | CONSTRUCTION PRINCIPLE	Tube fabric
4 | GEOMETRY	Three-dimensional
5 | MATERIAL	Polyester-laminated aluminum tape fabric
a) Procedural principle	Tubular fabric in which conductive aluminum ribbons are woven. Under pressure load, the hose is compressed and acts as a condenser. This produces a voltage change that can be correlated with the pressure load.
b) Schematic sketch	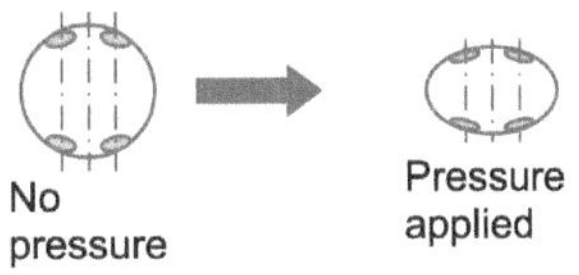
c) Known/possible field of application	Measuring changes in pressure load, e.g., decubitus prohylaxis or fall prevention; improvement of ergonomics.
d) Possible sensor variants	Temporal resolution subject to sensor design.
e) Opportunities and challenges	+ Tubular shape allows even compression under strain without the risk of the conductive belts shifting against each other + Production in one weaving process possible − Correlation of position/load and measuring signal for each position of the ligament tissue to be re-determined
I MATERIAL PROPERTIES	Operating range -20 to +50 °C
II ENERGY SUPPLY	Electric current
III RESOLUTION	250 ms
IV SENSITIVITY	Changes in capacitance of 2%
V MEASUREMENT RANGE	Pressure up to 150 kg
VI TRL	6–8

No pressure → Pressure applied

Figure 16. Description of the 3D tubular fabric in the catalog.

3.4. Outlook: Classified Knowledge within Platform-Based Smart Textile Development

This catalog, with its abstract solution description on the one hand and concrete information on areas of application on the other, enables developers to quickly and purposefully check which textile-based state-of-the-art sensors are suitable for a particular problem. This covers the critical part of "textile-based sensor technology" in the development of Smart Textiles. The catalog is an important starting point for an integrated product and process development of Smart Textiles and can be expanded in the future. The addition of new developments resulting from particular research work is necessary, e.g., to cover improvements in robustness towards broader usage requirements or the compliance with standards and test methods, which are still to be defined for Smart Textiles [17]. Furthermore, a reference to textile-integrated and adapted sensor technology, other functional components such as actuators, energy supply units and data transmission systems, as well as materials and production technologies for textile-electronic integration is promising (Figure 17).

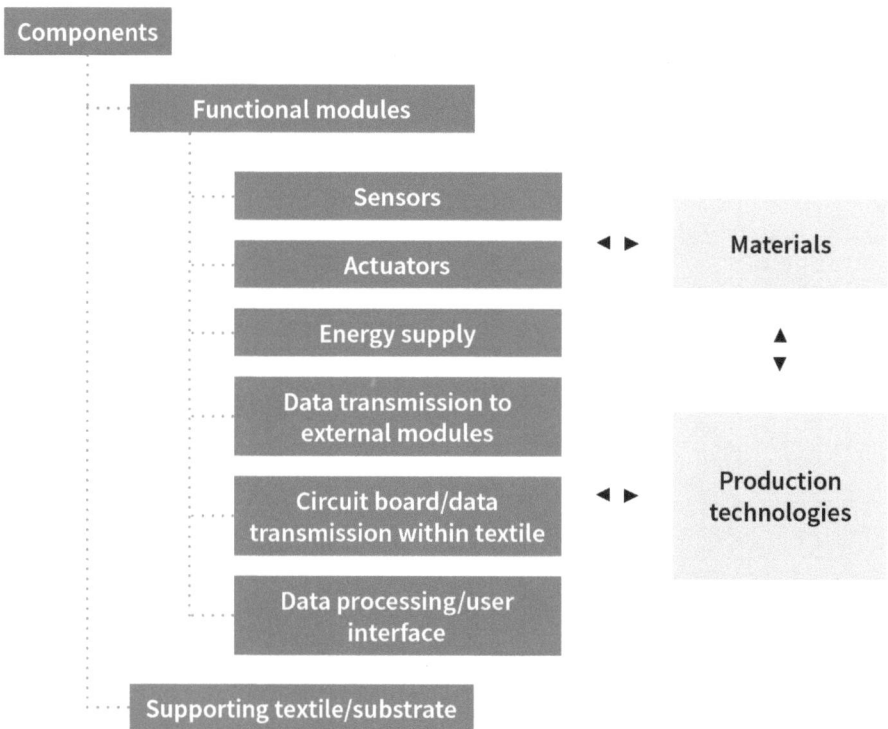

Figure 17. Modular structure of Smart Textiles [30].

Since there are many interdependencies between the choice of textile material and the properties of the functional components, and thus also with regard to possible processes, a structured information base for simplifying the Smart Textile development process is also helpful here. A platform-based information database can be implemented to reflect the complexity of mutual dependencies and the continuous further development of the state of the art technologies. Complex data structures and rule-based dependencies can be modeled and kept up-to-date in terms of content via an open, adaptable format. The project GeniusTex (2018, funded by the German Federal Ministry of Economic Affairs and Energy) will work towards this by implementing the collaboration platform GeniusTex (www.geniustex.net, [31]) with such a structured information base or language at its core (see Chapter 8).

PART IV
by Inga Gehrke and Vadim Tenner

Production Technologies for Electronic Textiles

The "combination" of electronics and textiles can be interpreted in several ways. A definition that concerns a structure of different integration stages is given below. Subsequently, the most important manufacturing processes and application examples are explained.

4.1. Integration Levels of Electronic Textiles

Intelligent textile systems differ in the extent to which their electrical components are integrated [15]. Table 2 distinguishes between three integration levels of Smart Textiles.

- Textile-adapted

This is the simplest variant of integration. Textile and electronics are separated (i.e., the textile is purely a shell). Example: An MP3 player is stored in a specially designed pocket in the clothing and cables are routed through eyelets and channels to the hood (Table 2, top).

- Textile-integrated

With textile-integrated Smart Textiles, individual functions are already completely mapped in textiles (i.e., produced by electrically conductive fiber materials and textile manufacturing processes). These include conductor tracks, heating loops and surfaces, resistors, capacitors and switches. The textile covers between 0 and 100% of the electronic function, creating an interface between the textile and the electronics. This can be bridged with various contacting methods. Most applications are currently at this level of integration.

- Textile-based

The electronic function is 100% covered by the textile. When considering an intelligent textile material, this can be the realization of conductor paths and sensors made of conductive yarns, piezoelectric fibers for energy generation or polymer optical fibers for light transmission.

Table 2. Integration levels of Smart Textiles [15].

Integration Level	Examples	
Textile-adapted	ScotteVest	Connection of the textile and the electrical components, e.g., via sewn-in pockets or Velcro fasteners.
Textile-integrated		Electrical components are integrated in the textile, e.g., using conductive yarns. Today's state-of-the-art.
Textile-based	Embroidered electrode pad	Textiles themselves take over the tasks of conventional hard electronic components, e.g., piezoelectric fiber, fluorescent fiber, field-effect transistor.

4.2. Textile Surface Processing for the Integration of Sensors and Conductive Tracks

If conductive yarns are processed with textile surface processes, conductive tracks, sensors and actuators can be produced. These "electrode pads" are suitable, for example, for measuring vital functions close to the body, muscle stimulation, monitoring functional components and as flexible operating elements, displays or heating elements. They have been the subject of research since the 2000s and have already been implemented in numerous prototypes (e.g., [19,32–35]).

4.2.1. Knitting

Knitted electrodes made of silver-coated polyamide fibers have already been used to monitor heart rhythm, respiration and bioimpedance (composition of body tissue). Footfalls and Heartbeats (UK) Limited (Nottingham, UK) commercially offers knitted sensors made of stainless-steel-coated polymer yarns (Figure 18). Resistances in the range of 5–5000 Ω cm^{-2} can be achieved, so that pressure sensors and electrodes can be implemented for various applications [36].

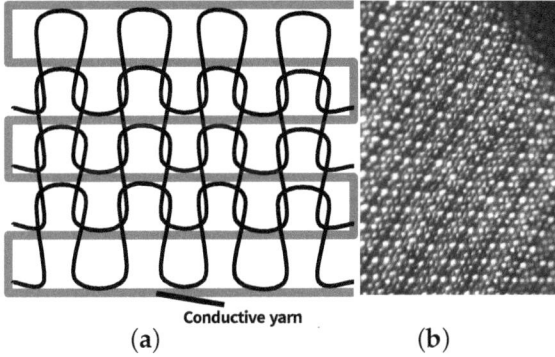

Conductive yarn
(a)   (b)

Figure 18. Schematic sketch (**a**) and photograph (**b**) of a knitted pressure sensor. Image courtesy of Sean Malyon, Footfalls and Heartbeats (UK) Limited.

In weft knits, the distance between conductors is limited to approximately 500 μm due to the size of the knit loops [37]. Li et al. used the intarsia knitting technique to produce circuits from copper fibers coated with polyurethane using a flatbed knitting machine (Figure 19) [38]. The weft knit is flexible and stretchable with only 1% change in electrical resistance after 1,000,000 stretching cycles with 20% maximum stretch. After 30 washing cycles for delicates at 30 °C, 16% of the samples showed a change in resistance. The weft knit was used in a demonstrator in a protective vest to measure load and strain [38]. Even though flatbed knitting can be automated, its reproducibility (see washability) and accuracy are still too low to knit circuits on an industrial scale.

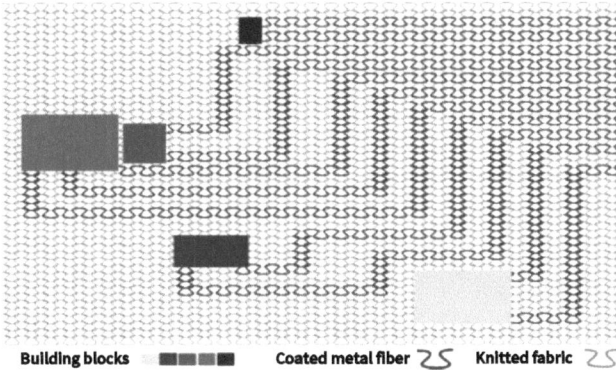

Building blocks ▀▀▀▀   Coated metal fiber ⌇   Knitted fabric ⌇

Figure 19. Circuit constructed using the intarsia knitting technique [38].

4.2.2. Weaving

Conductive yarns are used directly as warp threads in the weaving process and, depending on the weaving process used, form conductor paths. Conventional weaving allows only linear traversing webs, while the webs in Jacquard or open-reed weaving can also be shifted sideways to the production direction. Electrodes and sensors can also be manufactured in this way as shown in Figure 20.

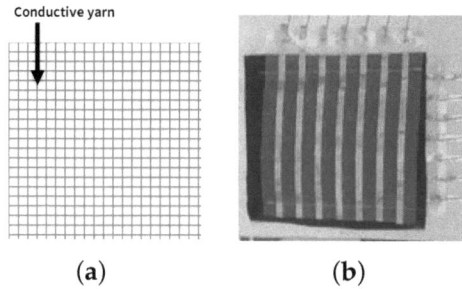

(a)   (b)

Figure 20. Schematic sketch and example of woven Smart Textiles. (**a**) Woven surface structure; (**b**) Woven pressure sensor.

In order to produce circuits in fabrics, further steps need to be followed after weaving with conductive yarns. For example, Locher et al. describe a method in which the insulating layer of a fabric made of polyester and insulated copper tracks is first removed with a laser at the desired points and then the conductor tracks are separated from the rest of the grid [39]. The cross points of warp- and weft-yarn are joined with a conductive adhesive and finally sealed with epoxy (Figure 21). Although circuit layouts can be realized in this way with a finer resolution (conductor spacing of 150 μm) than via knitting, the process is difficult to industrialize due to the additional steps and the layout can be designed less freely than in the case of embroidery or knitting [39].

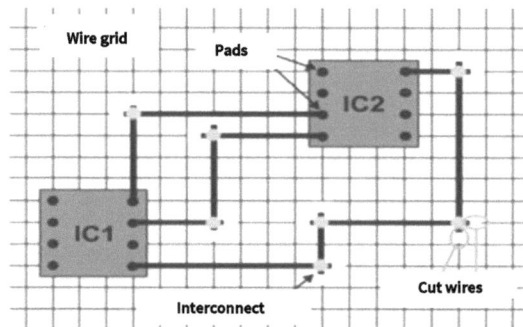

Figure 21. Woven circuit based on Locher et al. [39].

4.2.3. Spacing Textiles

As an alternative to flat fabrics, spacing textiles can also be used as sensors. A spacing textile consists of two fabric layers connected by pile yarns. In the KostBar project, a conductive aluminum-polyester band was inserted into a spacing textile as a pressure sensor, thereby functionalizing a demonstrator for evacuation mats in hospital beds [28]. Spacer weft- or warp-knits made of conductive yarns can also be used as resistive pressure sensors, in which pressure-induced contact between the upper and lower textile surfaces produces a change in resistance. Alternatively, intelligent materials can be introduced into the spatial structure. An example of this is the integration of fiber-based OLEDs into a spacer warp-knit as part of the "TexOLED" research project [40]. As shown in Figure 22, the light guide is inserted between the pile yarns, which provides the advantage of avoiding mechanical stresses such as bending in the sensitive yarn.

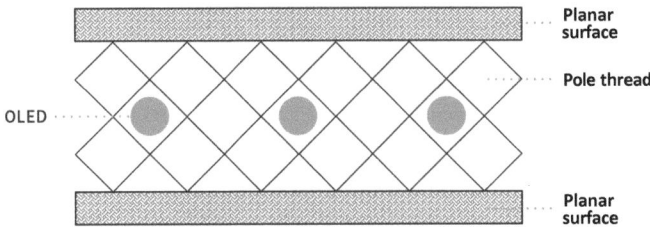

Figure 22. Principle for the installation of organic light-emitting diodes (OLEDs) in spacer textiles and prototypes (according to Linz [40]).

4.3. Subsequent Integration of Conductive Tracks and Sensors into the Textile Surface

In addition to direct conductor path integration during surface production, textile surfaces can also be finished with functionalized yarns (embroidering) or coatings (printing). This chapter provides the reader with an overview of how conventional production technologies from the textile and electronics industry can be combined to realize electronic circuits on textile materials. Embroidering and printing technologies as well as the related curing process for the creation of conductive structures are described.

4.3.1. Embroidering

The embroidery methods of chain stitch, standard, and Tailored Fiber Placement (TFP) embroidery are currently defined in the literature. With embroidery technology, flexible conductor track layouts can be realized from conductive yarns or even metal wires on textiles. In the TFP process, a laying thread is positioned with high precision on the textile substrate by means of an upper and a lower thread. This technology was initially developed for the production of fiber composites. If an electrically conductive yarn is used as the laying thread (Tailored Wire Placement), conductor paths can be implemented in this way. The TFP method is well suited for processing metallic threads [41,42]. The function of the TFP method can be seen in Figure 23.

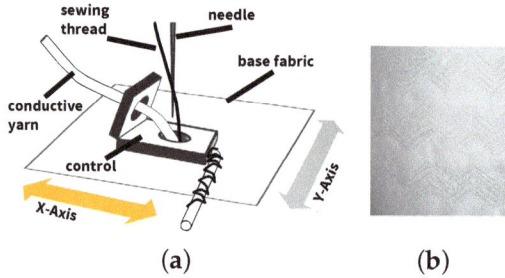

Figure 23. (a) Tailored Fiber Placement (TFP) method [41]; and (b) embroidered tracks [42].

Textile electronic circuits can be realized based on embroidery technology. For instance, a project by the research institute Textilforschungsinstitut Thüringen-Vogtland e.V. (Greiz, Germany) demonstrated embroidery technology of electrical connections between electronic components based on ELITEX conductive polyamide thread covered with silver, the embroidery technology of insulated conductor crossings and the contactability [43].

With double lockstitch embroidery, sensors, conductor paths, heating wires, etc., can be flexibly applied to textile surfaces. Depending on whether one or both of the upper and lower threads is conductive, single- or double-sided electrode pads can be produced (Figure 24) [44]. Wang et al. have realized radiofrequency (RF) antennas with metal-polymer fibers using embroidery, achieving signal strength only 1dB below conventional copper RF antennas. However, the durability and washability need to be tested [45].

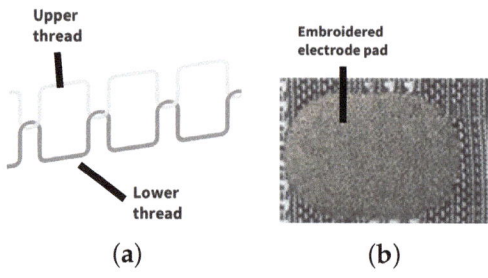

Figure 24. Schematic sketch of double lockstitch embroidery (a) and an electrode pad realized with it (b).

This technique is also known as "e-broidery". Under the trademark "e-broidery", Forster Rohner manufactures sensors as well as light-emitting textiles embroidered with LEDs (Figure 25) [44,46].

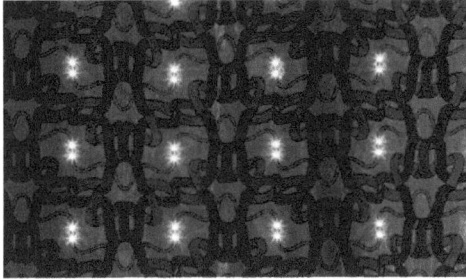

Figure 25. Embroidered light-emitting textiles. Image courtesy of Forster Rohner AG, St. Gallen, Switzerland.

Moss-embroidered electrodes have the advantage that permanent body contact can be better achieved through their 3D structure (Figure 26). Additionally, the shape and volume of the electrodes, which can be flexibly adapted to body shape, can be measured. Demonstrators for monitoring brain currents (electroencephalography [EEG] baseball cap), heart rate (electrocardiography [ECG] T-shirt) and fluid balance have been realized and successfully tested [47].

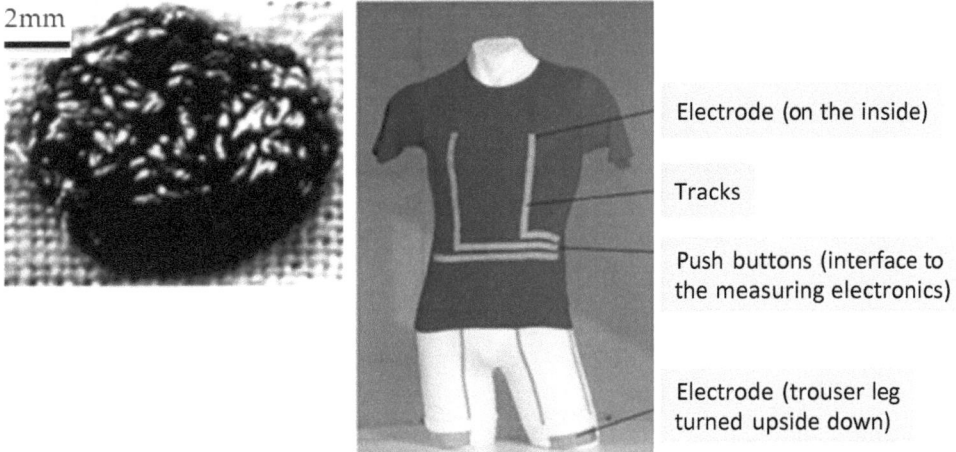

Electrode (on the inside)

Tracks

Push buttons (interface to the measuring electronics)

Electrode (trouser leg turned upside down)

Figure 26. Moss-embroidered electrode [47]. Clothing with embroidered tracks (**a**) and electrodes (**b**).

Microelectronic components can be contacted on the embroidered circuit using various methods. In addition to gluing and soldering, flexible circuits can be contacted directly with the embroidery thread (Figure 27a).

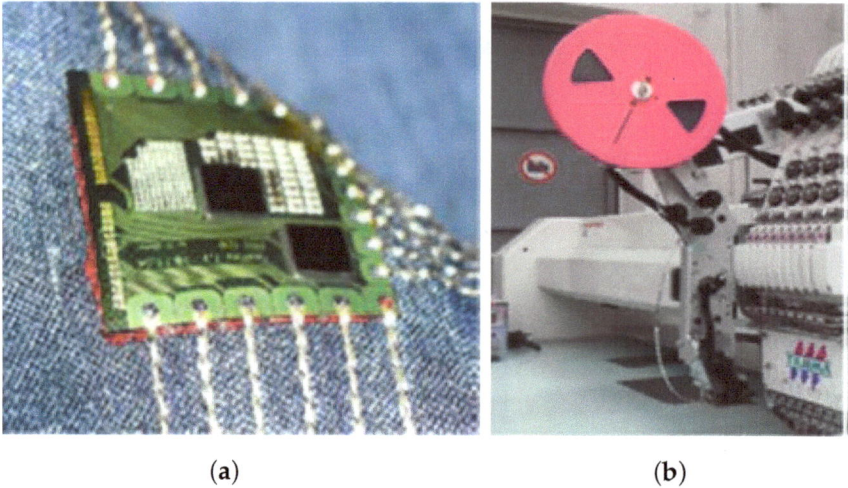

(a)   (b)

Figure 27. Flexible circuits contacted with embroidery thread (**a**) and sequin feeder on an embroidery machine (**b**) [43,48].

The research institute Textilforschungsinstitut Thüringen-Vogtland e.V. (Greiz, Germany), in cooperation with Tajima (Nagoya, Japan), has developed special sequins equipped with conductive structures and Surface Mount Device (SMD) components such as LEDs [43]. These "Functional Sequin Devices" can be directly applied and contacted via feed devices of the embroidery machine (Figure 27b).

Embroidery technology is a stable process for the integration of sensors and conductor track structures. The resolution is limited by the stitch size. On the other hand, the temperature load for the textile is low compared to the thermal load imposed by the electronic assembly processes (e.g., soldering, welding, curing).

4.3.2. Printed Circuit Boards on Textiles

Due to the possibility of higher resolutions of the conductive paths and therefore the possibility to integrate SMDs with much smaller dimensions, different printing technologies suitable for textile substrates are presented in the following.

4.3.2.1. Screen or Stencil Printing

While the previously mentioned production technologies use conductive yarns to achieve intelligent properties, Smart Textile applications involving printing on textiles are presented below.

Special inks allow the implementation of intelligent functions in textiles at high resolution. Washing resistance and susceptibility to cracking under mechanical stress are the greatest challenges in functional printing on textiles. Researchers at the University of Tokyo have developed a new type of conductive ink with high conductivity, mechanical strength and ease of use. The conductivity of an elastic conductor at an elongation of 0% is a maximum of 738 S cm^{-1}, and is a minimum of 182 S cm^{-1} at an elongation of 215% (cf. copper, 58–104 S cm^{-1} [49]). The components of the ink are silver flakes, fluorine rubber and surfactants. The fluorine surfactants arrange the conductive network in the conductor in such a way that high conductivity and ductility are achieved. The functionality of an organic transistor matrix stretched by 110% and that of an EMG sensor printed on textile have been proven [50]. Figure 28 shows the printed textile-integrated EMG sensor.

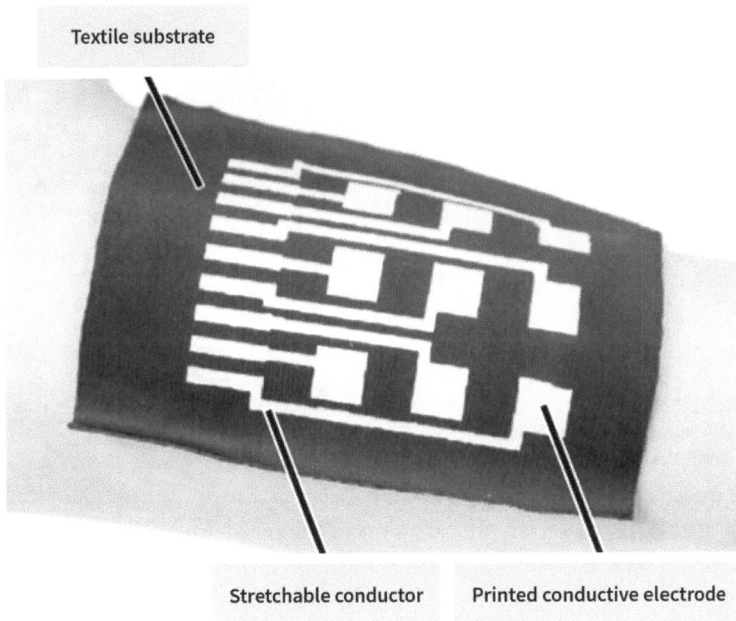

Figure 28. Printed textile-integrated EMG sensor [50].

The printing of classical color samples on textiles can be realized by different processes. These include rouleaux printing, flat stencil printing and rotary stencil printing. In rouleaux printing, the pattern to be printed is engraved on rollers which transfer the pattern to the textile. Screen-printing methods can be used to print conductive paths on a wide variety of materials, from textiles to foils and ceramics [51,52]. In flat stencil printing, also known as screen printing, the pattern is applied to a flat stencil. Printing is carried out for one stencil sheet after the other ("sheet to sheet", see Figure 29a).

The print paste is pressed through the permeable screen of the stencil onto the textile in the desired pattern. In places where the textile is not to be printed on, the stencil for the printing paste is impermeable. In rotary stencil printing, the stencil pattern is applied to a roller which prints the paste onto the textile ("roll to roll", see Figure 29b). In contrast to flat stencil printing, this process can then be carried out continuously [53,54].

The screen-printing process requires a subsequent curing of the printed textile in an oven, which is essential to maintain high conductivity and fix the printed material to the substrate (see Chapter 4.3.3) [55].

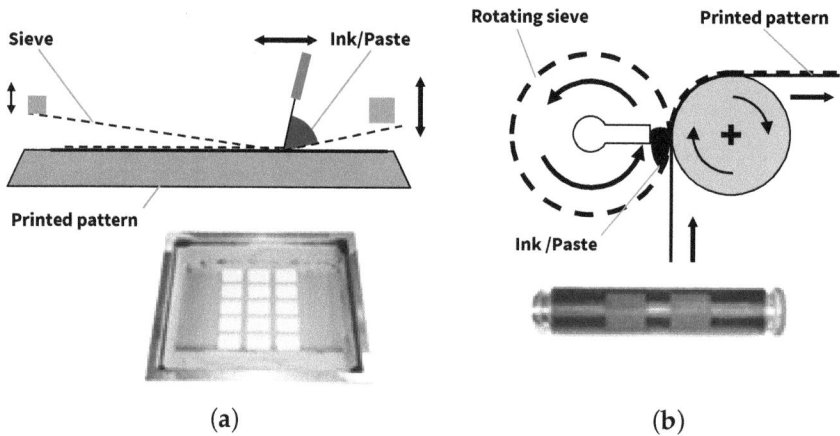

Figure 29. Process sequence for screen printing (**a**) and rotary printing (**b**).

4.3.2.2. Inkjet Printing

Inkjet printing is an alternative technique that does not require the creation of a stencil. It is a digital printing process that is used by most commercial paper printers. A digital image is processed by the printer and the print paste is applied to the carrier textile in small drops through a nozzle without touching the textile. This means that any pattern can be printed depending on the resolution of the printer. Only the processability of the printing paste for the printer and the printability of the textile are important [56,57].

When printing on textile, there are a few hurdles that have to be considered [57]:

• The uneven, unsmooth surfaces of textiles makes the uniform printing of conductive pastes difficult.
• Textiles are stretchy and flexible and should also be washable and breathable when used in clothing. This leads to an extraordinary load on the printed electronics.
• Any physical contact with the print imposes special demands on the print paste, which must therefore be skin-friendly.

4.3.2.3. The CREATIF Printer

In addition to conventional printing technologies, the so-called "CREATIF printer", named after the CREATIF research project, was developed by the Institut für Textiltechnik (ITA) of the RWTH Aachen University, Germany, the School of Electronics and Computer Science, University of Southampton, United Kingdom, and industry partners [58]. It is a digital printer equipped with print heads for functional pastes, an inkjet head and corresponding drying units. It prints on textiles with electrically conductive, thermochromic, luminescent, piezoresistive and many other pastes, which realize the smart functions of the textile. The components of conductive pastes are usually silver flakes, fluorinated rubber and fluorinated surfactants. Due to the high elasticity of the pastes, the high conductivity is maintained even when the material is stretched three times.

The printing is performed in layers. The DuPont 5025 paste (DuPont, Wilmington, Delaware, United States) is used for the production of the conductor paths. In the first layer, the non-intersecting tracks are printed, and in the following layer, a dielectric (insulation bridge) for the future intersections of the tracks is printed. This process is repeated until the complete electrical circuit has been realized. The operating principle of the CREATIF printer can be seen in Figure 30.

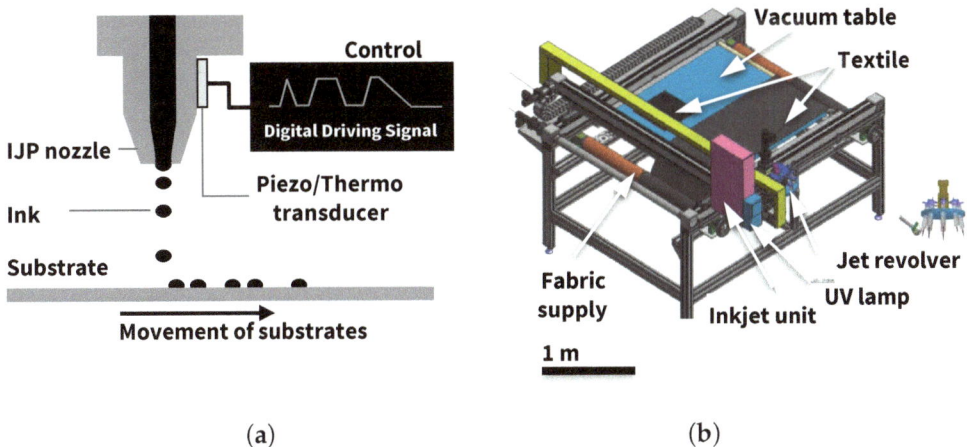

(a)   (b)

Figure 30. Procedural principle in functional printing (**a**) and CREATIF printer (**b**). IJP: ink jet printing.

At the ITA of the RWTH Aachen University, production concepts for Smart Textile products are developed and optimized continuously. By using the CREATIF printer, the printed circuit board is printed on a PVC-coated woven fabric as shown in Figure 31. Crossing conductive tracks can be bridged by printing a dielectric material in a multi-layer structure.

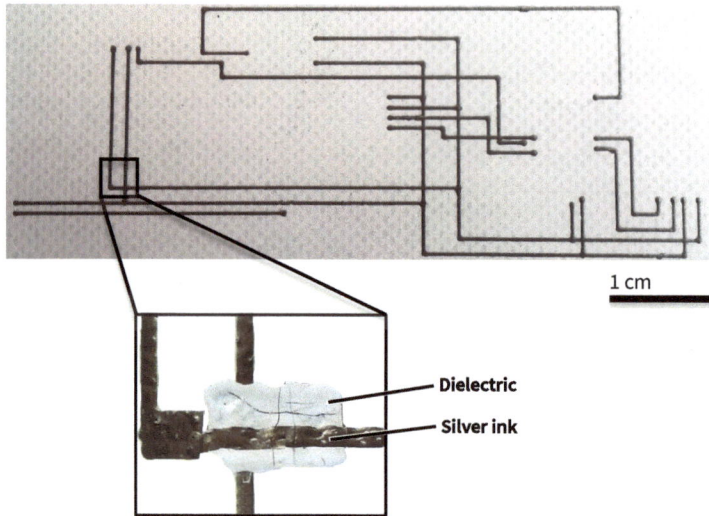

Figure 31. Printed circuit with insulation medium at crossover point [59].

The measured resistivity, ϱ, of the conductive ink, "DuPont 5025 Silver Conductor", in this case is $133\ \Omega \cdot mm^2\ m^{-1}$ (cf. silver, $\varrho = 0.015\ \Omega \cdot mm^2\ m^{-1}$ [60]).

4.3.3. Curing

In order to achieve high conductivity, the conductive tracks printed on the textile must be cured. Curing parameters can differ depending on the ink used. The samples created at the ITA of the RWTH Aachen University were inserted into a reflow oven at $130\ °C$ for 15 min. After repeating the curing process four times, a decrease in the electrical resistance was observed (Figure 32).

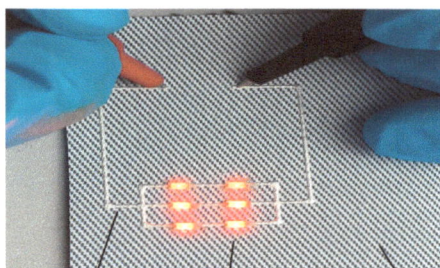

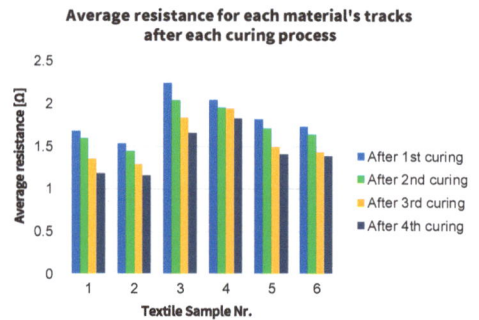

(a)   (b)

Figure 32. Electric circuit (**a**) and conductivity of the tracks after washing (**b**).

4.4. Contacting Method between Textile and Electronics

At present, there are four common methods of contacting electrical components with conductor paths. Table 3 gives an overview of the procedures [43].

Table 3. Common types of processes for contacting textiles and electronics [43].

Procedure	Contacting Method	Suitability
1	Directly via soldering	(+)
2	With electrically conductive adhesives	(-) Risk of short-circuits if adhesive penetrates the textile in case of narrow conductor spacings (<2 mm)
3	First soldered onto an interposer, which is stitched on with electrically conductive threads	(+)
4	Direct stitching-on of component connections	(-) Not suitable for small SMD components

Table 4 gives an overview of alternative contacting possibilities between electronic SMDs and textiles.

Table 4. Contacting possibilities between Surface Mount Device (SMD) components and textiles [43,61,62].

The soldering of electronics onto textiles has been declared to be a promising process, even under high mechanical stress [43]. Reflow- and laser-soldering are two methods that have been tried and tested in industry for contacting SMD components.

4.4.1. Manual Soldering

Manual soldering is the simplest soft-soldering process (<450 °C) used to create soldered joints. The soldering process takes place in three steps: first, the conductor paths are thermally or mechanically cut through; second, one end of the trace is cut; and finally, the free end of the trace is threaded through the pin hole on the underside of the SMD component and the solder point is set (see Figure 33).

Figure 33. Example of the manual soldering process.

The manual joining of small electronic components onto textiles involves numerous potential sources of error. For instance, the textile can easily burn through when touching the soldering iron, and the position accuracy of the joint is comparatively low. Automated processes can increase production speed by a factor of 30.

For example, Molla et al. contacted SMD-LEDs on stitched circuits using manual reflow soldering. Fourteen-hour mechanical wear tests showed a 3% failure rate [63].

4.4.2. Laser Soldering

Laser soldering is particularly suitable for the production of Smart Textiles, since the focused laser beam causes only a spatially limited and short-term thermal load on the textile [43]. The principle of laser soldering is shown in Figure 33.

4.5. Coating to Improve the Washability of Textile-Integrated Electronics

Removing the electronics before washing is generally not a desirable solution for Smart Textiles. However, a market-ready solution for this challenge has not yet been found. Different coating and encapsulation techniques are explored. For instance, the department of textiles of Ghent University researched the improvement of washability of integrated textiles with SMDs [64]. The research aimed to improve the washability by using a protective polyurethane layer for covering conductive tracks, printed on different textiles—Cotton (CO), Viscose (CV), Polyamid (PA) and Polyester (PES). Washing trials according to ISO 6330:2000 in a domestic washing machine at 40 ± 3 °C showed that about half of the used samples lost their conductivity after 20 washing cycles. Conductive tracks were printed with the commercial inks Electrodag PF 410 and 5025 (Henkel AG & Company, KGaA, Düsseldorf, Germany) [64]. Furthermore, Molla et al. improved the durability of reflow solder joints on stitched circuit traces using polymer tapes as encapsulation. The best samples could withstand up to 1000 min of washing and drying [65].

Additional research regarding washable Smart Textiles is conducted at the ITA of the RWTH Aachen University. In order to test the washability of coated SMDs integrated into a textile and their connection strength, 30 batches, each including eight conductive tracks, were tested for their electrical conductivity after washing. The used SMDs varied between LEDs (NEVARK 5988210107F) and resistors with different sizes. The conductive tracks were screen-printed with silver paste. To create a protection layer against environmental influences, the SMDs were coated with silicone.

The results show that all of the 150 SMDs survived the first washing cycle after being coated with silicone. After repeating the washing trial 20 times, nearly every LED survived. These results provide a clear indication of a strong connection between the SMDs and the conductive tracks, even after repeated mechanical stress.

4.6. Approaches to Automating Smart Textile Production

Exemplary processes for the automated, cost-effective production of Smart Textiles are presented below (Figure 34). To this end, various research institutes are pushing ahead with machine developments in order to map the entire process chain of electronic integration in textiles in a multifunctional device. The production steps (based on [43]) are listed chronologically below:

1. Embroidery or printing of the conductive paths in the mounted textile;
2. Dispensing solder paste or conductive glue on the SMD footprints;
3. Automated transfer of SMDs from the production platform and positioning on the textile with the aid of a vacuum gripper;
4. Contacting by soldering of the SMDs with the conductor tracks.

The multifunction device shown in Figure 35 contains the following components:

- Semiconductor laser soldering device from MiLaSys Technologies GmbH (Holzgerlingen, Germany);
- Dispenser, also from MiLaSys technologies GmbH;
- Round rotatable production platform for electronic components;
- Vacuum gripper for holding and positioning electronic components on the embroidery base;
- Charge-coupled device (CCD) camera for monitoring the soldering process.

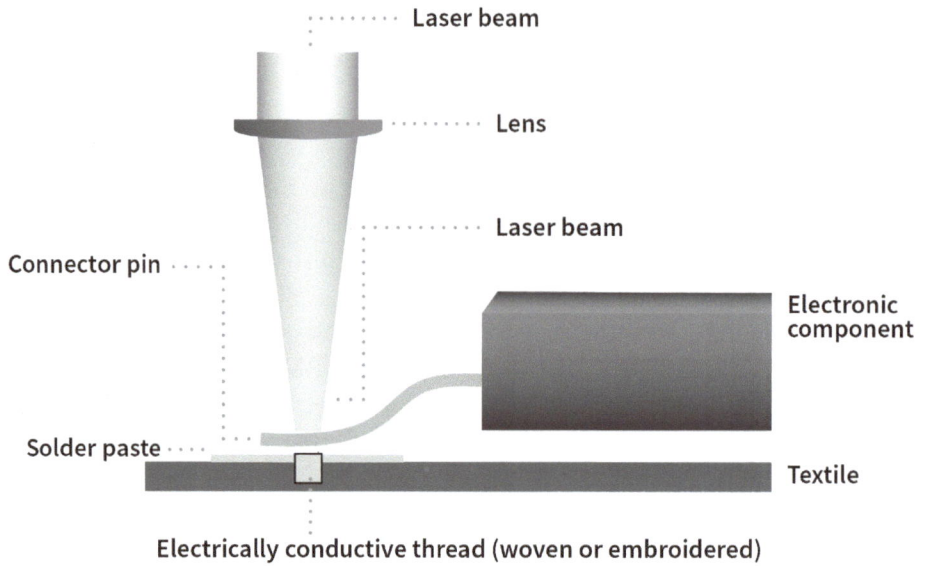

Figure 34. Principle of laser soldering (according to Reference [43]).

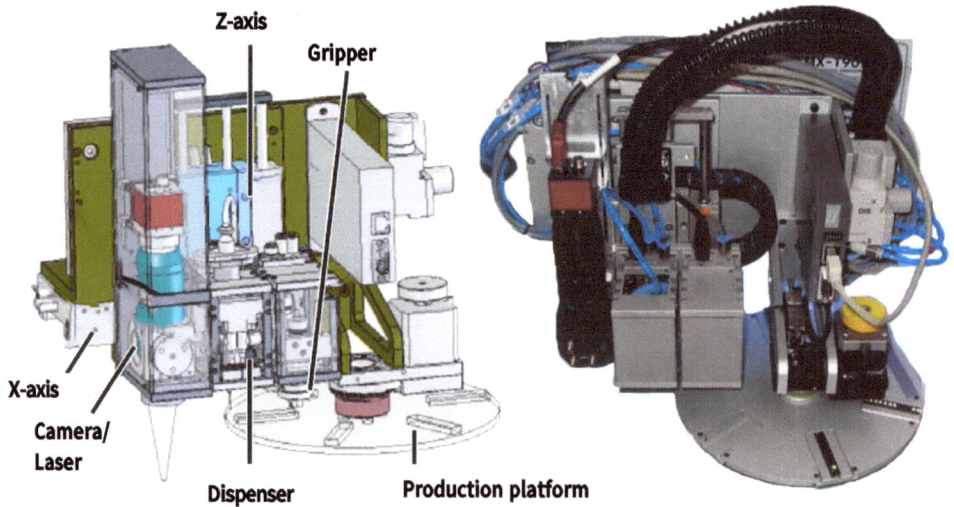

Figure 35. Multifunction device [43].

The installation of the multifunction device takes place directly at the fourth function head of the TCWM TRIPPLE-QUATTRO embroidery machine (Tajima GmbH, Winterlingen, Germany).

In the first step, the embroidery machine creates conductor tracks on the embroidery ground. Then, the multifunction device dispenser places a solder paste or adhesive at the end of the tracks. The vacuum gripper sucks the SMDs out of the round production platform and positions them at the conductor track ends of the embroidery ground. Finally, the SMDs are soldered on using a soldering robot.

A matrix of RGB-LEDs (KIRRON lightning components GmbH & Co KG, Korntal-Münchingen, Germany) was constructed based on the presented process. The functional model is shown in Figure 36 [43].

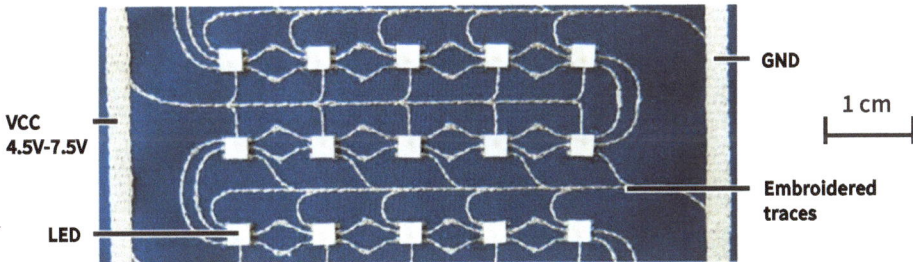

Figure 36. Functional pattern of an LED matrix [43]. GND: ground; VCC: voltage at the collector.

4.7. Automation Concepts

The ITA of the RWTH Aachen University has developed complete process chains for the automated production of Smart Textiles [59]. In the following, the production steps required for the efficient manufacture of a wristband equipped with sensors are described (Figure 37).

In clocked production, each layer of printed circuit paths requires a drying time of at least 10 min at 120 °C, depending on the printing paste used. The less layers that need to be printed, the shorter the production time. This aspect must be taken into account when designing the trace.

The prerequisites are given to apply the mechanical and/or thermal separation process for the whole textile. The adoption of existing separation technologies in the textile industry is conceivable.

The goal of the automated joining process is to minimize production time and guarantee reproducible results. The joining is carried out on a P30 Pick & Place machine (Mechatronic Systems, Tegernheim, Germany). The advantages of the machine are its two optical measuring systems and graphical interface, which facilitate the placement of the SMDs on the upper and lower side of the substrate. Table 5 provides the technical data of the machine.

1. Conductor track printing on the textile with the help of the CreatiF printer

2. Cutting the textile to size with a thermal/mechanical tool

3. Assembly of the textile with adhesive and electronics using the Pick & Place machine P30

4. Adhesive curing of the assembled textile in a reflow oven

5. Quality control of the contacting between electronics and textile with the help of a multimeter

6. Application of non-heat-resistant electronics, a protective foam bag and a finished layer

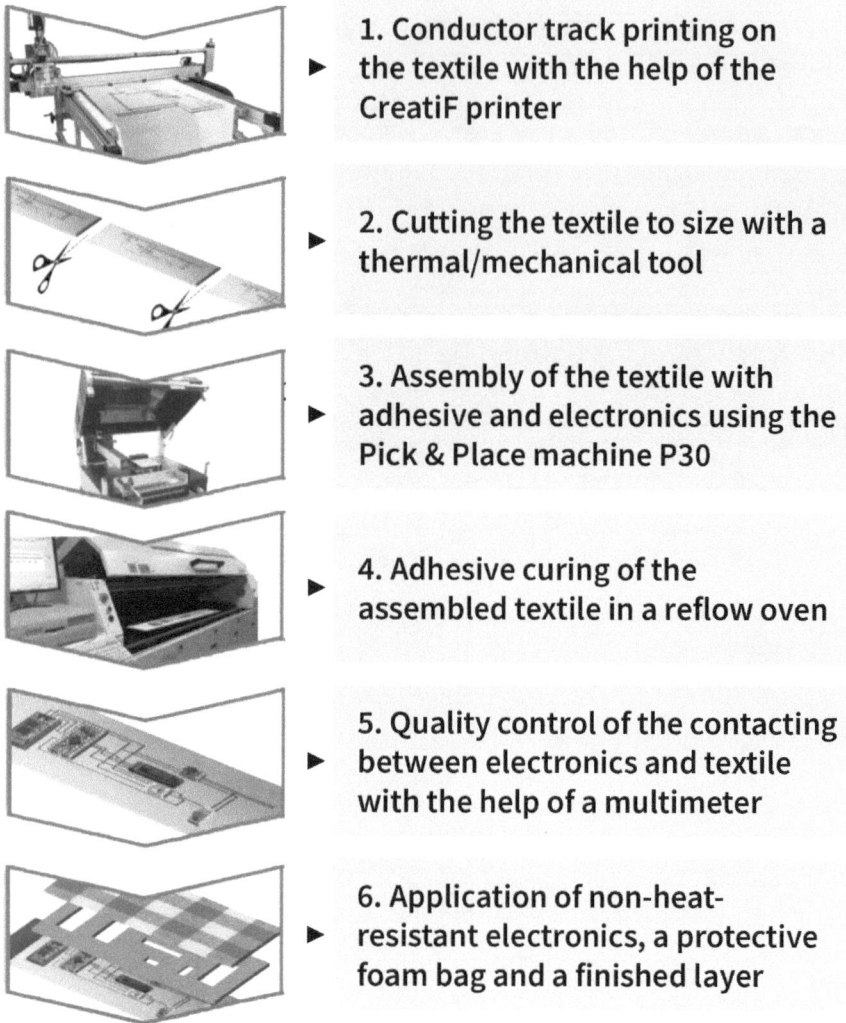

Figure 37. Production chain of a Smart Wristband at the ITA [59].

Table 5. Technical data of the P30 Pick & Place machine.

Designation	Characteristics
Work area	500 × 480 mm
SMD size	<40 × 40 mm
Application speed	1200 parts/hour
Adhesive dosing speed	6000 points/hour
Precision	0.4 mm

The Cadsoft Eagle PCB Design software (Version 9, Autodesk, Inc.), a design software for electronics boards, is used to provide the necessary Gerber and Pick-and-Place files. In the first step, a library is created for each component to be added to. This contains the size and position of the connection pins (footprints) of the respective component. Then, a Printed Circuit Board (PCB) is created with the aid the software program, based on the formed libraries. The PCB layout with square footprints is shown in Figure 38.

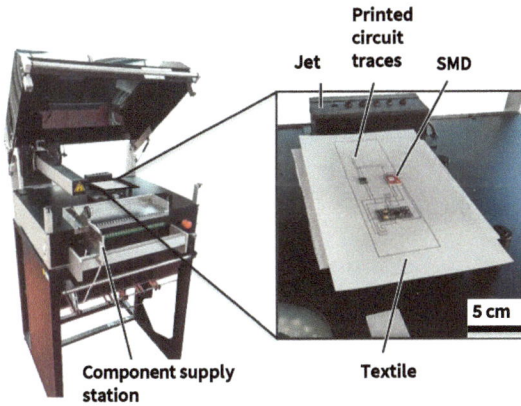

Figure 38. Pick & Place machine and automatically applied electronics (center) [59].

The electronics can be protected from mechanical influences by laminating on a foam cover layer. The end-product of an intelligent textile wristband is shown in Figure 39.

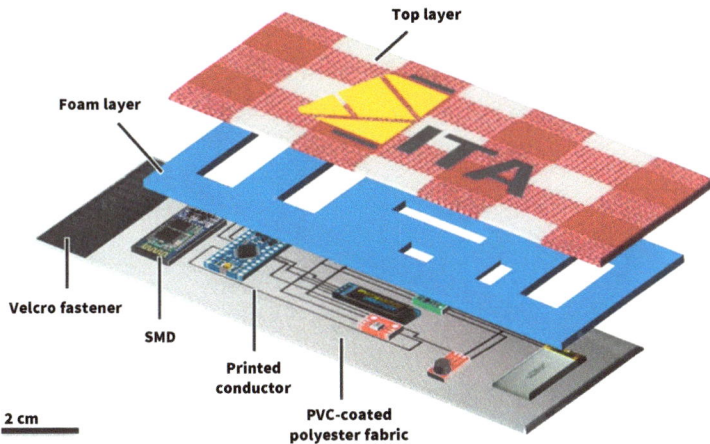

Figure 39. Model of the Smart Wristband [59].

PART V
by Vadim Tenner

Smart Textiles Product Concepts—
Design and Examples

There is a great deal of current research in the field of textile touchpads. As it occurs with commercially available touchpads, various electrical principles are used. As well as touchpads, these principles are also used for touch and pressure sensors. In particular, such sensors are increasingly used in medical technology [66–68]. In the following, this product type is used to describe design concepts and examples for a specific Smart Textile, building on the materials, sensors and production technologies introduced in the previous chapters.

5.1. Resistive Touchpads/Sensors

5.1.1. Sensomative

Sensomative (Rothenburg, Switzerland) is a start-up company, founded in 2015 in Switzerland, which produces textile pressure-measuring mats [69]. These are based on the same principle as resistive touchpads (see Figure 40). The sensor mass consists of two layers of conductive textile which are separated from each other by a spacer grid. By the construction of the pressure mat from many individual sensors and a suitable measuring algorithm, a pressure distribution can be represented with the sensomative mat. The mats are used, for example, to control the sitting posture of office chairs and wheelchairs in order to draw attention to posture errors and uneven loads [66,69] (see Figure 41).

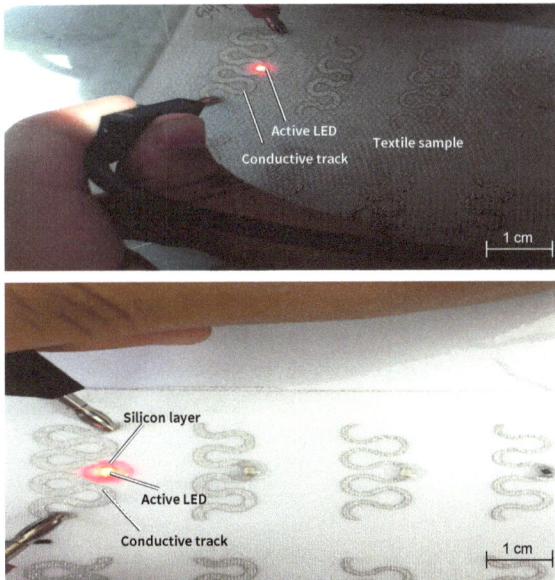

Figure 40. LEDs on printed conductive tracks without (**upper**) and with (**lower**) silicone coating.

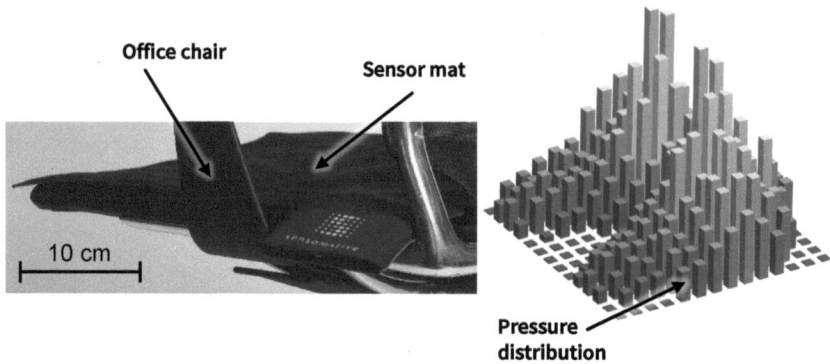

Figure 41. Sensomative pressure-measurement mat and signal representation [69]. Image courtesy of Sensomative.

5.1.2. Fabri Touch

Fabri Touch is a joint project of the Media Computing Group of the Department of Computer Science of the RWTH Aachen University and Smart Wearables Ltd. (Sofia, Bulgaria). The project deals with the integration of a resistive touchpad on the thigh-front of a pair of trousers. For this purpose, a layer structure consisting of a piezoresistive film, a spacer grid and an electrically conductive fabric is used. The assembly is completed with non-conductive textiles at the top and bottom. The piezoresistive film changes its resistance depending on the applied force. If a voltage is applied to the conductive fabric, the distance between the piezoresistive film and the fabric changes. The distance is ensured by a grid. The change in distance causes a current to flow from the fabric to the film. By means of voltmeters at all corners of the film, the position and pressure of the touch can be determined with the help of the appropriate resistance. Within the project, a prototype was produced and its suitability for the recognition of gestures and the control of a mouse pointer was examined. That prototype has registered gestures, and could be used to operate a mouse pointer. However, not all gestures could be registered equally well: experiments showed that horizontal gestures were less precise than vertical gestures, and additionally operation on the thigh was less stable than on a firm support [70].

5.2. Capacitive Touchpads and Sensors

5.2.1. Amotape Pressure Sensor

The Amotape Pressure Sensor of AMOHR Technische Textilien GmbH (Wuppertal, Germany) is a sensor in the form of a tape based on the principle of electrical capacitance. The sensor consists of a hose whose two layers approach each other under load, which results in a measurable change in capacitance (Figure 42).

The tape can be used in the care sector to protect against bedsores or falls from bed [67,68].

Figure 42. Tubular pressure sensor. Image courtesy of AMOHR GmbH [68].

5.2.2. Google Jacquard

The Internet company Google published its Google Jacquard concept in 2015 at "Google I/O", its annual developer conference. Within the framework of this project, a yarn that can be used for electrical applications was developed that can be woven like normal yarns. Copper wire is coated with polyurethane and then yarn is braided around it. This gives the multi-component yarn a purely textile structure on the outside so that the yarn does not stand out optically. The polyurethane layer protects the copper core from the influence of chemicals and high temperatures and further prevents skin contact. Yarns produced in this way should have mechanical properties comparable to those of conventional yarns and can be processed in the standard weaving process. This creates a textile in which the electrical components are not visible from the outside. By connecting suitable electronics, the woven yarns are transformed into a textile touchpad based on the principle of self-capacity. As the hand approaches, the electric field around the yarns changes. This change can be evaluated by the microelectronics used and translated into gestures. Google's concept has been applied in a denim jacket by Levi Strauss & Co. The touch area is integrated into the fabric on the sleeve. The microcontroller can be removed to wash the jacket. All other components of the jacket are washable [71,72].

PART VI
by Volker Lutz and Inga Gehrke

Summary and Outlook

Despite significant advances in both hardware and software technology and user interaction design, Smart Textiles have not taken off yet beyond the prototype stage. One of the major reasons for this is the complexity of products, technologies and businesses, which has prevented Smart Textiles from becoming market-ready products.

This book offers the possibility to structure the predominant complexity of Smart Textile products and to assist in the design and selection of required technologies. Its structured description of potential materials and technologies forms a basis to efficiently network with the necessary stakeholders of the textile, electronics, software, design and service industries.

This book provides a basis that can be used by all players, especially for the description of individual textile and textile-related components. Additionally, its structure provides a basis for the targeted and coordinated further development of materials, technologies and processes in order to bring the textile and electrical engineering sectors much closer together.

A directly usable tool is the sensor catalog presented in this book. The following steps are necessary to ensure that further developments lead to direct updating of the catalog:

- Creation of an open-source database that can be updated and reviewed by all potential stakeholders.
- Extension of the database by elementary knowledge of marketable production technologies.
- Extension of the database with design guides and business models.
- Generation of real work results and business-to-business (B2B) relationships to create marketable products.
- Use of the database in future interdisciplinary training concepts of necessary qualifications for Smart Textiles.

To overcome the known market barriers, especially for small and medium enterprises, the project GeniusTex (2018, funded by the Federal Ministry of Economic Affairs and Energy, Germany) will create cooperation and collaboration opportunities to develop Smart Textile products, services and business models (see Figure 43).

Based on an interactive innovation platform, GeniusTex (www.geniustex.net) will enable a B2B business model between Smart Textile developers, producers and users. Within the project, a methodology for process design to integrate textiles and electronic components into Smart Textiles will be developed. As part of that process, the sensor catalog introduced in this book is one technical outcome of this development platform.

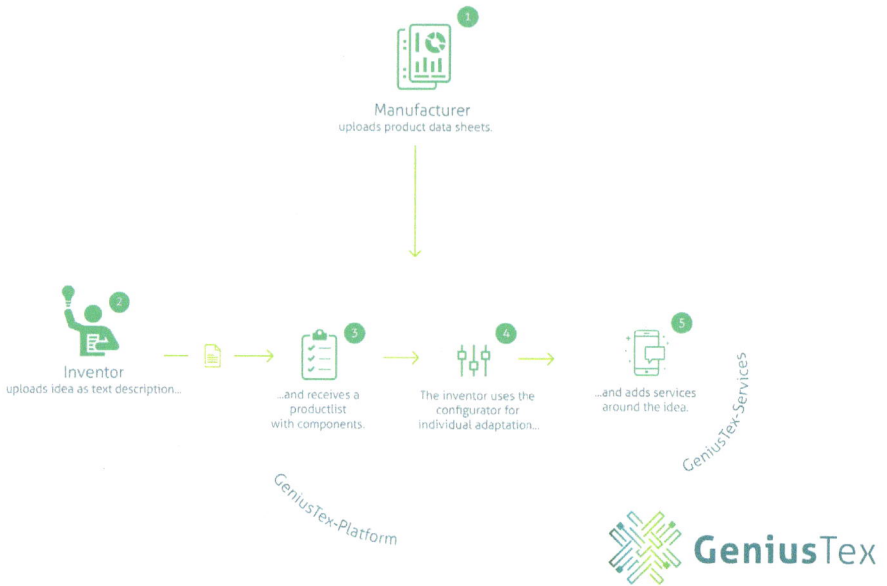

Figure 43. GeniusTex Smart Textiles platform and its collaborators. SDK: software development kit.

The innovation platform will be open to all players in Smart Textiles (textiles, electronics, designers, software application developers, end-users) and their contributions (e.g., by software development kit). The platform also aims to overcome the strong segmentation in the textile industry in particular. Since almost every production step is done by individual parties, an enhanced communication and material flow is necessary. It is expected that only fast and standardized communication between each step or new multi-stage production steps will solve this challenging problem.

A common language for Smart Textile components, building on the modularization described in Part II (Figure 17), will be defined in order to structure offerings on the platform and to connect ideas and partners. The platform will have an international setup (e.g., USA, South Korea) to ensure that both platform architecture and structured language are globally accessible. Together with the feedback of end-users, it will generate the opportunity to consider the market needs at an early stage of product development.

To support early-stage developments, the GeniusTex software development kit (SDK) will help to connect inhomogeneous sensors and to develop intelligent Smart Textile services. A web-based graphical editor will be created to simplify the selection, crosslinking and definition of trigger events, as well as the selection of actuators.

Besides collaboration platforms like GeniusTex, it is crucial that product development is not in particular based on technology push. Smart Textiles need the consideration of any kind of technology that is feasible under the consideration of application requirements such as functionality, acceptance and usability. The integration of technology and textiles should be taken into account whenever the unique properties of textiles are essential for the desired use case, however not for the sake of textile integration itself.

Parallel to further developments in materials and process technology research, the creation of adjacent infrastructures for Smart Textiles is necessary. These developments include:

- Standardization: required to facilitate stakeholder interactions, e.g., creating interfaces between electronics and textiles, contacting, data standards, testing standards, etc.
- Concepts for the user-accepted, sustainable and economic integration of energy supplies.
- Increased and application-specific robustness of the components used, but also defined mechanical or chemical stress, e.g., washing along the entire product life cycle.
- Business models for the entire Smart Textiles system. It can be assumed that the mostly digital services associated with Smart Textiles will account for a significant proportion of the future economic performance of these products.
- Regulations and laws that support the intended usage scenario and prevent misuse.

So far, established companies from the textile, electronics, software, etc., sectors have not been able to build sustainable Smart Textile business models without expanding their existing competencies. This means that small companies or start-ups have an especially high chance of success in the future Smart Textiles market, due to the fact that small enterprises are more agile in adopting other technologies. Nevertheless, both small and established companies face the same challenge of moving from demonstrators and prototypes to economically successful products. For this purpose, scalable production technologies for Smart Textiles must be generated in the participating industries. These technologies need to represent the step from small series to mass production without suffering a technological or economic break.

Future Smart Textiles products will also require the creativity and courage of designers and engineers to create new products. In order to move more quickly from the creative product-development process to the industrially manufacturable product in the future, flexible, modular design systems are required that combine the basic technologies. By flexibly combining technologies such as printing, embroidery, pick and place, or cutting in one manufacturing system, designers and engineers can develop and manufacture products from prototypes to small batches. The use of such systems should not be restricted to technology experts. That means that the basic technologies for functionalizing textiles need to be easily accessible to a wide range of potential operators. These flexible manufacturing systems also represent a toolbox for new business models in contract manufacturing or finishing, e.g., conventional textile products could be refined and finished by small enterprises with smart functions.

Author Contributions: The authors contributing to the individual parts of this work are named at the beginning of each section. Within the sections, the authors have contributed equally to this publication.

Acknowledgments: The publishing of this work has been supported by funding from the Federal Ministry for Economic Affairs and Energy of Germany for the GeniusTex project.

Conflicts of Interest: The authors declare no conflict of interest. The founding sponsors had no role in the design of the study; in the collection, analyses, or interpretation of data; in the writing of the manuscript, and in the decision to publish the results.

REFERENCES

1. Hayward, J. *E-textiles 2018–2028*; Technologies, Markets and Players: Cambridge, MA, USA, 2018.

2. CEN Europäisches Kommittee für Normung. *Textilien und textile Produkte—Intelligente Textilien—Definitionen, Klassifizierung, Anwendungen und Normungsbedarf*; Beuth Verlag GmbH: Berlin, Germany, 2012.

3. Gartner. Gartner Hype Cycle. Available online: https://www.gartner.com/en/research/methodologies/gartner-hype-cycle (accessed on 1 November 2018).

4. Steinmann, W.; Schwarz, A.; Jungbecker, N.B.; Gries, T. *Faserstofftabelle elektrisch leitfähige fasern*; Shaker: Aachen, Germany, 2014.

5. Cork, C.R. Conductive fibres for electronic textiles: An overview. In *Electronic Textiles: Smart Fabrics and Wearable Technology*; Woodhead Publishing Series in Textiles 166; Dias, T., Ed.; Elsevier: Amsterdam, The Netherlands, 2015; pp. 3–20.

6. Qu, H.; Skorobogatiy, M. Conductive polymer yarns for electronic textiles. In *Electronic Textiles: Smart Fabrics and Wearable Technology*; Woodhead Publishing Series in Textiles 166; Dias, T., Ed.; Elsevier: Amsterdam, The Netherlands, 2015; pp. 21–53.

7. Miao, M. Carbon nanotube yarns for electronic textiles. In *Electronic Textiles: Smart Fabrics and Wearable Technology*; Woodhead Publishing Series in Textiles 166; Dias, T., Ed.; Elsevier: Amsterdam, The Netherlands, 2015; pp. 55–72.

8. Bredas, J.L.; Marder, S.R.; Salaneck, W.R. Alan J. Heeger, Alan G. MacDiarmid, and Hideki Shirakawa. *Macromolecules [Online]* 2002, *35*(4), 1137–1139.

9. Posudievsky, O.Y.; Konoshchuk, N.V.; Shkavro, A.G.; Koshechko, V.G.; Pokhodenko, V.D. Structure and electronic properties of poly(3,4-ethylenedioxythiophene) poly(styrene sulfonate) prepared under ultrasonic irradiation. *Synth. Met.* **2014**, *195*, 335–339. [CrossRef]

10. Worfolk, B.J.; Andrews, S.C.; Park, S.; Reinspach, J.; Liu, N.; Toney, M.F.; Mannsfeld, S.C.B.; Bao, Z. Ultrahigh electrical conductivity in solution-sheared polymeric transparent films. *Proc. Natl. Acad. Sci. USA* **2015**, *112*, 14138–14143. [CrossRef] [PubMed]

11. Ryan, J.D.; Mengistie, D.A.; Gabrielsson, R.; Lund, A.; Müller, C. Machine-Washable PEDOT:PSS Dyed Silk Yarns for Electronic Textiles. *ACS App. Mater. Interfaces* **2017**, *9*, 9045–9050. [CrossRef] [PubMed]

12. Pani, D.; Dessi, A.; Saenz-Cogollo, J.F.; Barabino, G.; Fraboni, B.; Bonfiglio, A. Fully Textile, PEDOT:PSS Based Electrodes for Wearable ECG Monitoring Systems. *IEEE Tran. Bio-Med. Eng.* **2016**, *63*, 540–549. [CrossRef] [PubMed]

13. Åkerfeldt, M.; Lund, A.; Walkenström, P. Textile sensing glove with piezoelectric PVDF fibers and printed electrodes of PEDOT. *Text. Res. J.* **2015**, *85*, 1789–1799. [CrossRef]

14. Bosowski, P.; Husemann, C.; Quadflieg, T.; Jockenhövel, S.; Gries, T. Classified Catalogue for Textile Based Sensors. *AST* **2012**, *80*, 142–151. [CrossRef]

15. Bosowski, P.; Hörr, M.; Mecnika, V.; Gries, T.; Jockenhövel, S. Design and manufacture of textile-based sensors. In *Electronic Textiles: Smart Fabrics and Wearable Technology*; Woodhead Publishing Series in Textiles 166; Dias, T., Ed.; Elsevier: Amsterdam, The Netherlands, 2015; pp. 75–107.

16. de Acutis, A.; de Rossi, D. e-Garments: Future as "Second Skin"? In *Smart Textiles: Fundamentals, Design, and Interaction*; Schneegass, S., Ed.; Springer: New York, NY, USA, 2017; pp. 383–396.

17. European Commission. *Smart Wearables Reflection and Orientation Paper: Digital Industry Competitive Electronics Industry*; Technical Report; European Commission: Brussels, Belgium, 2016. Available online: ec.europa.eu/newsroom/document.cfm?doc_id=40542 (accessed on 1 November 2018).

18. Mattmann, C.; Clemens, F.; Tröster, G. Sensor for measuring strain in textile. *Sensors* **2008**, *8*, 3719–3732. [CrossRef] [PubMed]

19. Suh, M. Wearable sensors for athletes. In *Electronic Textiles: Smart Fabrics and Wearable Technology*; Woodhead Publishing Series in Textiles 166; Dias, T., Ed.; Elsevier: Amsterdam, The Netherlands, 2015; pp. 257–273.

20. Nayak, R.; Wang, L.; Padhye, R. Electronic textiles for military personnel. In *Electronic Textiles: Smart Fabrics and Wearable Technology*; Woodhead Publishing Series in Textiles 166; Dias, T., Ed.; Elsevier: Amsterdam, The Netherlands, 2015; pp. 239–255.

21. Chen, D.; Lawo, M. Smart Textiles and Smart Personnel Protective Equipment. In *Smart Textiles: Fundamentals, Design, and Interaction*; Schneegass, S., Ed.; Springer: New York, NY, USA, 2017; pp. 333–357.

22. Lorussi, F.; Carbonaro, N.; de Rossi, D.; Tognetti, A. Strain- and Angular-Sensing Fabrics for Human Motion Analysis in Daily Life. In *Smart Textiles: Fundamentals, Design, and Interaction*; Schneegass, S., Ed.; Springer: New York, NY, USA, 2017; pp. 49–70.

23. Gries, T.; Jungbecker, N.B.; Leonhardt, S.; Röthlingshöfer, L.; Kim, S. *Textilintegriertes, intelligentes System zum Ernährungs- und Wasserhaushaltsmanagement—NutriWear*; Schlussbericht Rahmenprogramm "Mikrosysteme 2004–2009" des BMBF Förderkennzeichen 16FV3480; Leibniz Information Centre for Science and Technology University Library: Hannover, Germany, 2010.

24. Paradiso, R.; Loriga, G.; Taccini, N.; Gemignani, A.; Ghelarducci, B. WEALTHY-a wearable healthcare system: New frontier on e-textile. *J. Telecommun. Inf. Technol.* **2005**, *4*, 105–113.

25. Gries, T.; Veit, D.; Wulfhorst, B. *Textile Fertigungsverfahren: Eine Einführung*; Carl Hanser Verlag GmbH Co. KG: Munich, Germany, 2014.

26. Verein Deutscher Ingenieure. *VDI 2222 Blatt 2: Konstruktionsmethodik: Erstellung und Anwendung von Konstruktionskatalogen*; VDI-Verlag: Düsseldorf, Germany, 1982; pp. 1–13.

27. Mankins, J.C. *Technology Readiness Levels*; White Paper; Advanced Concepts Office, Office of Space Access and Technology, NASA (The National Aeronautics and Space Administration): Washington, DC, USA, 1995.

28. Bosowski-Schönberg, P. *Kostbar—kontinuierliche Fertigung von 3D-Smart Textiles-Bandgewebe am Beispiel funktionalisierter Evakuierungsmatten*; Schlussbericht BMBF Förderkennzeichen 16SV58; Institut für Textiltechnik der RWTH Aachen University: Aachen, Germany, 2015.

29. Bosowski-Schönberg, P. Entwicklungsmethodik für textile Überwachungssysteme am Beispiel der Drucksensorik. Ph. D. Dissertation, Technische Hochschule, Aachen, Germany, 2016.

30. Gehrke, I.; Schmelzeisen, D.; Gries, T. Platform-based product development for Smart Textiles. Presented at Aachen Innovation Platform, Aachen, Germany, 25 September 2018.

31. Gehrke, I.; Schmelzeisen, D. GeniusTex: Welcome to the GeniusTex Project (Project Homepage). Available online: https://geniustex.net/ (accessed on 6 December 2018).

32. Pfeiffer, M.; Rohs, M. Haptic Feedback for Wearables and Textiles Based on Electrical Muscle Stimulation. In *Smart Textiles: Fundamentals, Design, and Interaction*; Schneegass, S., Ed.; Springer: New York, NY, USA, 2017; pp. 103–137.

33. Mbise, E.; Dias, T.; Hurley, W. Design and manufacture of heated textiles. In *Electronic Textiles: Smart Fabrics and Wearable Technology*; Woodhead Publishing Series in Textiles 166; Dias, T., Ed.; Elsevier: Amsterdam, The Netherlands, 2015; pp. 117–132.

34. Jungbecker, N.B. Gestaltung und Charakterisierung textiler Elektroden zur Überwachung von Körperfunktionen. Ph. D. Dissertation, Technische Hochschule, Aachen, Germany, 2011.

35. Peiris, R.L. Integrated Non-light-Emissive Animatable Textile Displays. In *Smart Textiles: Fundamentals, Design, and Interaction*; Schneegass, S., Ed.; Springer: New York, NY, USA, 2017; pp. 71–101.

36. Hayward, J. *E-textiles 2016–2026*; Company profiles 1601A: Cambridge, MA, USA, 2016.

37. Varga, M. Electronics Integration. In *Smart Textiles: Fundamentals, Design, and Interaction*; Schneegass, S., Ed.; Springer: New York, NY, USA, 2017; pp. 161–184.

38. Li, Q.; Tao, X.M. *Three-Dimensionally Deformable, Highly Stretchable, Permeable, Durable and Washable Fabric Circuit Boards*; The Royal Society: Cambridge, UK, 2014; p. 470.

39. Locher, I.; Troster, G. Fundamental building blocks for circuits on textiles. *IEEE Trans. Adv. Packag.* **2007**, *30*, 541–550. [CrossRef]

40. Linz, T. *TexOLED: Textilintegrierte und textilbasierte LEDs und OLEDs, Entwicklung neuer Technologien zur Erzeugung textiler Flächen und Fäden mit hoher Leuchtdichte*; Schlussbericht Förderkennzeichen 16SV3451; Leibniz Information Centre for Science and Technology University Library: Berlin, Germany, 2010.

41. Mecnika, V.; Scheulen, K.; Anderson, C.F.; Hörr, M.; Breckenfelder, C. Joining technologies for electronic textiles. In *Electronic Textiles: Smart Fabrics and Wearable Technology*; Woodhead Publishing Series in Textiles 166; Dias, T., Ed.; Elsevier: Amsterdam, The Netherlands, 2015; pp. 133–153.

42. Eichhoff, J.; Hehl, A.; Jockenhövel, S. Textile fabrication technologies for embedding electronic functions into fibres, yarns and fabrics. *Multidiscip. Know-How Smart-Text. Dev.* **2013**, *7*, 191–226.

43. Textilforschungsinstitut Thüringen-Vogtland e.V. *Verfahren zur effizienten Realisierung von stoff- und formschlüssigen elektrischen Verbindungen bei der Fertigung von Smart Textiles*; Schlussbericht 17055 BR: Greiz, Germany, 2013.

44. Hörr, M.; Jockenhövel, S.; Gries, T. Sticken zur Funktionalisierung von Textilien. *Magazine for Textile Decoration and Promotion* **2015**, 4, 58–61.

45. Wang, Z.; Volakis, J.L.; Kiourti, A. Embroidered antennas for communication systems. In *Electronic Textiles: Smart Fabrics and Wearable Technology*; Woodhead Publishing Series in Textiles 166; Dias, T., Ed.; Elsevier: Amsterdam, The Netherlands, 2015; pp. 201–237.

46. Castano, L.M.; Flatau, A.B. Smart fabric sensors and e-textile technologies: A review. *Smart Mater. Struct.* **2014**, 23, 53001. [CrossRef]

47. Hörr, M.; Jockenhövel, S.; Gries, T. Textile Elektroden durch Moos- und Kettelstickerei. *Magazine for Textile Decoration and Promotion* **2015**, 3, 68–72.

48. Kallmayer, C. *TePat—Textile Mikrosystemtechnikplattform*; Schlussbericht Förderkennzeichen 16SV4055: Germany, 2012.

49. N.N. Elektrische Leitfähigkeit, Datenblatt, 2016. Available online: https://www.isabellenhuette.de/fileadmin/Daten/Praezisionslegierungen/Datenblaetter_Widerstand/E_KUPFER.pdf (accessed on 6 December 2018).

50. Matsuhisa, N.; Kaltenbrunner, M.; Yokota, T. Printable elastic conductors with a high conductivity for electronic textile applications. *Nature Communications* **2015**, 6, 7461. [CrossRef] [PubMed]

51. Orthmann, K. *Kleben in der Elektronik*; Expert-Verlag: Renningen, Germany, 1995.

52. Kim, Y.; Kim, H.; Yoo, H.-J. Electrical characterization of screen-printed circuits on the fabric. *IEEE Trans. Adv. Packag.* **2010**, 33, 196–205.

53. Gries, T.; Veit, D.; Wulfhorst, B. *Textile Fertigungsverfahren—Eine Einführung*; Carl Hanser Verlag: München, Germany, 2015.

54. Stolz, S. Siebdruck von elektrisch leitfähigen Keramiken zur Entwicklung heizbarer keramischer Mikrokomponenten. Ph.D. Dissertation, Albert-Ludwigs-Universität, Freiburg im Breisgau, Germany, 2002.

55. Locher, I.; Tröster, G. Screen-printed Textile Transmission Lines. *Text. Res. J.* **2007**, 77, 837–842. [CrossRef]

56. O Ecotextiles. Digital Printing. Available online: https://oecotextiles.wordpress.com/tag/inkjet-printer/ (accessed on 7 September 2017).

57. Torah, R.; Wie, Y.; Li, Y.; Yang, K.; Beeby, S.; Tudo, J. Printed Textile-Based Electronic Devices. In *Handbook of Smart Textiles*; Tao, X., Ed.; Springer Science + Business Media: Singapore, 2015; pp. 653–687.

58. Tudor, J.; Torah, R. *CreaTiF: Digital creative tools for digital printing of smart fabrics*; Förderkennzeichnen CP-FP-INFSO-FP7-610414; University of Southampton: Southampton, UK, 2016.

59. Tenner, V. Holistic approach towards the development of customised Smart Textile products—Material development, production technologies and services. Presented at the ITMC2017—International Conference on Intelligent Textiles and Mass Customisation, Ghent, Belgium, October 2017.

60. Frederikse, H.P.R. Techniques for Materials Characterization. Experimental Techniques Used to Determine the Composition, Structure, and Energy States of Solids and Liquids. In *CRC Handbook of Chemistry and Physics*, 90th ed.; Lide, D.R., Ed.; CRC Press/Taylor and Francis: Boca Raton, FL, USA, pp. 41–42.

61. Monfeld, C. Smart textiles - textiles with enhanced functionality, 2015. Available online: www.smarttextile.de (accessed on 7 September 2017).

62. Arum, L. Lilypad holder. Available online: https://www.thingiverse.com/thing:13198 (accessed on 6 December 2018).

63. Molla, M.T.I.; Goodman, S.; Schleif, N.; Berglund, M.E.; Zacharias, C.; Compton, C.; Dunne, L.E. *Surface-Mount Manufacturing for E-Textile Circuits*; ACM: New York, NY, USA, 2017.

64. Kazani, I.; Hertleer, C.; de Mey, G.; Schwarz, A.; Guxho, G.; van Langenhove, L. Electrical conductive textiles obtained by screen printing. *Fibres Text. Eastern Eur.* **2012**, *20*, 57–63.

65. Molla, M.T.I.; Compton, C.; Dunne, L.E. *Launderability of Surface-Insulated Cut and Sew E-Textiles*; ACM: New York, NY, USA, 2018.

66. Plüss, S. (CTO and Co-Founder SensoMATive). Introduction to the sensor mat SensoMATive. Personal communication at TechTextil, Frankfurt am Main, Germany, 2017.

67. Amohr Technische Textilien GmbH. Produkte—Elektrisch Leitfähige Bänder, Elastische Kabel und Textile Sensoren. Available online: http://www.amohr.com/produkte/leitfaehige-baender-textilien/ (accessed on 7 September 2017).

68. Amohr Technische Textilien GmbH. Flyer, 2017.

69. Sensomative. Available online: http://sensomative.com/de/ (accessed on 1 June 2017).

70. Heller, F.; Ivanov, S.; Wacharamanotham, C.; Borchers, J. FabriTouch: Exploring Flexible Touch Input on Textiles. In Proceedings of the 2014 ACM International Symposium on Wearable Computers, Seattle, WA, USA, 13–17 September 2014; pp. 59–62.

71. Poupyrev, I.; Gong, N.; Fukuhara, S.; Karagozler, M. Project Jacquard: Interactive Digital Textiles at Scale. In Proceedings of the 2016 CHI Conference on Human Factors in Computing Systems, San José, CA, USA, 7–12 May 2016; pp. 4216–4227.

72. Google Atap. Project Jacquard. Available online: https://atap.google.com/jacquard/ (accessed on 7 September 2017).

APPENDIX

by Inga Gehrke and Patrycja Bosowski-Schoenberg;
Design and and Illustrations of Catalog by Jan Serode

Classified Catalog of Textile-Based Sensors for Developing Smart Textiles

	AGROTECH	
	LIQUID LEVEL INDICATOR	**HUMIDITY SENSOR**
| SENSOR TYPE	Chemical	Chemical
| MEASURAND	Electromagnetic light spectrum (refractive index fiber core/fiber cladding)	Electric current
| CONSTRUCTION PRINCIPLE	Fiber	Weft knit
| GEOMETRY	Linear	Planar
| MATERIAL	Glass	Electrically conductive yarn
Procedural principle	The liquid filling level is determined by measuring a proportion of the light coupled out in the optical waveguide correlating to the refractive index of the liquid medium. The propagation of the light within the fiber is influenced by the surrounding liquid medium. The geometry of the fiber section determines the cancellation of the total reflection of light waves. [1]	Knitted fabric with a basic weft knit which contains at least one thread made of a material which changes its electrical resistance when affected by moisture. The weft knit is equipped with an integrated moisture sensor, consisting of at least two electrodes arranged at a distance, which are electrically connected to each other in case of moisture. [2]
Schematic sketch	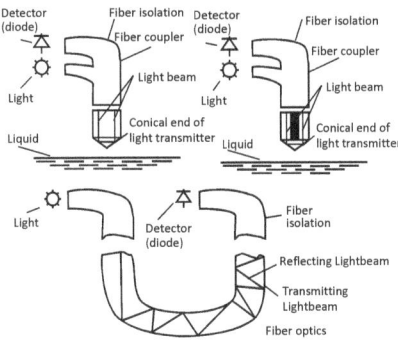	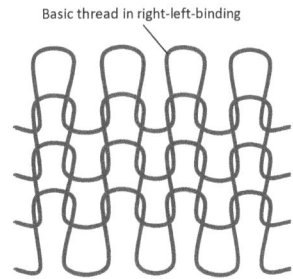
Basic thread in right-left-binding
Known/possible fields of application	Possible applications in chemical process engineering, e.g., determination of the acid concentration via the refractive index.	Woven fabrics in which electrically well conducting and electrically not well conducting threads are alternately woven with each other.
Electrical means of connection in the form of terminals, plug-in connection parts.
Possible sensor variants	Variation of the fiber geometry: Conical ends of the optical waveguide and miniature prisms cancel out the total reflection for all parts of the light waves. A fiber bent into a U-shape and provided without cladding glass only picks up a certain part of the light waves. A fiber-optic refractometer sensor of high sensitivity is also suitable as a temperature sensor.	Electrical means of connection can be connected to the monitoring station via textile conductors. The textile behavior ensures that the joint is extremely flexible and elastic.
Opportunities and challenges	+ Possible corrosion of sensor
+ Restriction of the measuring range
+ Long-term stability

− Danger of contamination of the sensor | + Integration of the sensor directly into the garment, with no external application necessary |
MATERIAL PROPERTIES		
ENERGY SUPPLY	None	Electric current
| RESOLUTION		
| MATERIAL		
MATERIAL PROPERTIES		
| TRL	9	6–8

	OPTOELECTRONIC SENSOR	MOISTURE- AND CHEMICAL-SENSITIVE SENSOR THREAD
1 | SENSOR TYPE	Chemical	Mechanical, chemical
2 | MEASURAND	Electromagnetic light spectrum	Visual assessment
3 | CONSTRUCTION PRINCIPLE	Fiber	n/a
4 | GEOMETRY	Linear	Linear
5 | MATERIAL	Plexiglas	n/a
a) Procedural principle	Detection of adhering liquid components in or on liquid-storing substances by detecting the change in the transmission of light in a light guide with the liquid component to be taken. [3]	Permanent identification of harmful environmental influences through the use of threads which change their shape, color or volume while absorbing liquid. The core yarn must be UV-resistant and clearly distinguished in color from the load-bearing tape. For the sheath fibers of the yarn, a material must be selected which is changed in shape, color or structure by UV radiation. [4]
b) Schematic sketch	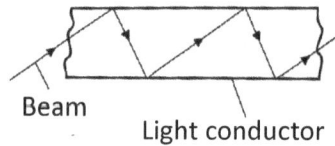 Beam / Light conductor	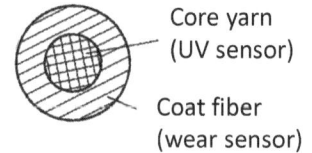 Core yarn (UV sensor) / Coat fiber (wear sensor)
c) Known/possible field of application	Detection of liquid content of soils, textiles or granulates. Monitoring tasks, for example in landfills.	A friction-spun sensor thread represents a combination of abrasion and UV sensor.
d) Possible sensor variants	Cost-effective	Decrease in abrasion resistance with increasing exposure to UV radiation.
e) Opportunities and challenges		
I MATERIAL PROPERTIES		
II ENERGY SUPPLY	Light	None
III RESOLUTION		
IV SENSITIVITY		
V MEASUREMENT RANGE		
VI TRL	9	9

	AGROTECH	
	WATER DETECTOR	**DETECTION MEANS**
SENSOR TYPE	Chemical	Chemical
MEASURAND	Visual assessment	Visual assessment
CONSTRUCTION PRINCIPLE	Textile tape, thread, thread bundle, textile fiber composite, fleece, paper, film, wire, warp knit	Fiber
GEOMETRY	Punctiform, linear, planar, voluminous	Linear, planar
MATERIAL	Cellulose, polyolefin, nylon, Nomex, Teflon, plastic, polyester, ceramic, metal, wool	Cellulose, plastic, glass, ceramics
Procedural principle	Textile probe with sufficiently large stored active substance depot, which on contact with the substance to be investigated causes a visual chemical change in the detector depending on the composition and movement of the analyte. The change occurs in the form of a substance solution, substance deposition or formation of a new substance at the detector itself. [5]	Detection of substances with shaped and unshaped detection means, containing fibers and/or adhesives which react to environmental influences via a color change as an indicator. [6]
Schematic sketch		
Known/possible field of application	Analysis of gas and water, and also soil and sediment, samples.	Analysis of water, soil and sediment samples for natural and artificial constituents including radioactive contaminants. Control measures in food and feed production. Production and monitoring of industrial products, including gases. Monitoring and control of industrial processes. Control measures in the nuclear sector.
Possible sensor variants	The resistance of the optically visually-recognizable color pattern of the detector to water with a different composition to that of the measuring point and the atmosphere, which is exposed to short-term effects, prevents falsification of the measurement.	Spatially and temporally seamless qualitative monitoring and documentation of processes possible.
Opportunities and challenges		– The detection medium can also be used to a limited extent as a filter for certain substances
MATERIAL PROPERTIES		
ENERGY SUPPLY	None	None
RESOLUTION	>1 h	
MATERIAL		
MATERIAL PROPERTIES		
TRL	9	9

	AGROTECH	
	FIBER-OPTIC SENSOR	**CARBON-FILLED CELLULOSE FIBER**
1 | SENSOR TYPE	Chemical	Mechanical
2 | MEASURAND	Electromagnetic light spectrum	Electric current
3 | CONSTRUCTION PRINCIPLE	Fiber	Fiber, filament, film
4 | GEOMETRY	Linear	Linear, planar
5 | MATERIAL	Cotton for protective vision, fluoride glass for light guide sheath and core	Polymer
a) Procedural principle	Fiber-optic sensor for detecting gaseous or liquid media, surrounded by an optical fiber sheath consisting of a fluoride glass of low chemical resistance to be detected on contact with the analyte. Decomposition of the sheath takes place within a characteristic chemically induced reaction time until the sensor responds as a function of the original thickness of the sheath, the temperature and the concentration of the attacking medium while maintaining the total reflection condition (lower refractive index of the sheath with respect to the optical fiber core). A textile, hygroscopic protective layer around the light-guide sheath increases the corrosive effect of the attacking medium on the light-guide sheath. [7]	Carbon-filled cellulose fiber. Detection of liquids or vapors via electrically conductive filaments from dry-wet spun cellulose dotted with charge carriers (graphite, carbon black, pigments with semiconducting layers, metallic fibers or carbon fibers) whose conductivity changes under tension/ pressure or with increasing moisture content. [8]
b) Schematic sketch	Fiber-optic core Fiber-optic sheath Gas- and liquid-permeable protective cover	Humidity Drag C-doped core Print Mantle
c) Known/possible field of application	Detection of gaseous and liquid media. Monitoring of electrical cables, lines and endangered installations, pipelines, equipment and buildings for the ingress of water, water vapor, acids, alkalis or other gases and liquids.	Detection of liquids or vapors.
d) Possible sensor variants	High mechanical strength.	Mechanically stable even at high temperatures, and sometimes even fire-retardant.
e) Opportunities and challenges	+ High response sensitivity, even to individual media only + Targeted analysis of individual specific substances with desired concentration content + Low manufacturing and general costs	− Increasing carbon-black content reduces substance strength, ductility and toughness − Doping with carbon-black influences the material viscosity to such an extent that the formation of stable threads is not possible at normal spinning speeds − If the doping with soot is too high, the electrical resistance increases disproportionately
I MATERIAL PROPERTIES	Light-guide sheath with lower refractive index than conductor core, light-guide sheath made of fluoride glass with lower hydrolytic resistance	
II ENERGY SUPPLY	None	Electric current
III RESOLUTION		
IV SENSITIVITY		
V MEASUREMENT RANGE		
VI TRL	6–8	6–8

AGROTECH

	PH SENSOR	FIBER-OPTIC PH SENSOR
SENSOR TYPE	Chemical	Chemical
MEASURAND	Visual assessment	Electromagnetic light spectrum
CONSTRUCTION PRINCIPLE		Fiber
GEOMETRY	Linear	Linear
MATERIAL		Polymer, glass
Procedural principle	Measurement of substance concentrations, which are not directly accessible spectroscopically, with a sensitive chemoreceptor. This receptor is a sensor, at the end of which a specific indicator (e.g., phenol red in polyacrylamide) is immobilized by which a change in pH is measured either in reflection or as fluorescence. [9]	Utilization of the light absorption dependent on the pH value of the surrounding medium in a fiber-optic probe consisting of a segment of a multimode optical fiber whose end forms the sensor head. In this area, both the coating and the cladding of the fiber are removed so that a sensitive layer of a copolymer with immobilized dye is polymerized onto the core. Electromagnetic radiation is guided in such a way that the light rays pass through the interface between the fiber core and the sensitive layer and are returned to the core by total reflection at the interface between the sensitive layer and the aqueous analyte. Wavelength-selective absorption occurs. [10]
Schematic sketch
Optical fiber
Optical fiber
Immobilized indicator
Permeable membrane
Cladding Coating pH-sensitive layer Mirror layer
Protective layer
Unmirrored face Fiber core Shaft Epoxy resin 6–20 mm
Known/possible field of application		Chemical-analytical measurements.
Possible sensor variants	Very accurate measurement of the pH value only achievable for very small ranges (approximately three pH units).	Low influence of the internal thickness on the sensor characteristic curve.
Opportunities and challenges		+ High long-term stability
+ High sensitivity
+ Damping arm
MATERIAL PROPERTIES		Six months service life
ENERGY SUPPLY	None	Light
RESOLUTION		
MATERIAL		
MATERIAL PROPERTIES	0.005 pH units	At 680 nm, 0.06 absorbance units per pH unit over the measuring range of four pH units
TRL	6–8	9

	AGROTECH	
	INTEGRATED OPTICAL FREQUENCY DOUBLER	**INTEGRATED OPTICAL RESONATOR**
1 | SENSOR TYPE	Thermal	Thermal
2 | MEASURAND	Electromagnetic light spectrum	Electromagnetic light spectrum
3 | CONSTRUCTION PRINCIPLE	Fiber-optic conductor	
4 | GEOMETRY	Linear	Linear
5 | MATERIAL		$LiNbO_3$
a) Procedural principle	Determination of absolute temperatures by means of optical frequency doubling, in which a specific light wavelength is required for a known temperature of the resonator in order to achieve a frequency conversion (phase-matching of fundamental and harmonic wave) with high efficiency. [1]	The temperature is changed by means of optical resonators integrated in LiNbO3 with a periodic characteristic curve. To be able to record the number of orders passed as a function of the direction of the phase (or temperature) change requires two signals phase-shifted by 90°. It is advantageous to use the output signals to arrive at an evaluation, which counts in each case with the zero crossing, and thus an independence from slow fluctuations of the light intensity is achieved. The phase modulation required for differentiation is achieved by frequency modulation of the laser light or by electro-optical modulation of the optical path length of the resonator. [1]
b) Schematic sketch	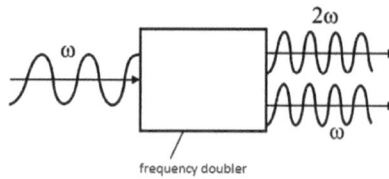 frequency doubler	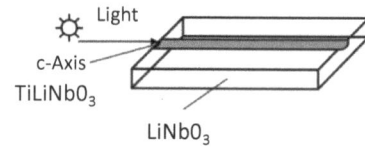
c) Known/possible field of application	Temperature monitoring of textile structures.	Temperature monitoring of textile structures.
d) Possible sensor variants	Particularly high efficiency.	The sensitivity of the temperature sensor can be determined in wide ranges by the length of the component and the wavelength of the light.
e) Opportunities and challenges	− The prerequisite for measurement is a tunable, coherent light source with enough power to operate the resonator.	+ Simple measuring system with high accuracy when supplying the resonator sensor element via a polarization-maintaining monomode fiber + Measurement of the smallest temperature changes possible due to the strong temperature dependence of the refractive index − Measurement of absolute temperatures not possible
I MATERIAL PROPERTIES		
II ENERGY SUPPLY	Electric current	None
III RESOLUTION		
IV SENSITIVITY		Sensitivity of 35 impulses/K, resolution of 29 impulses/K
V MEASUREMENT RANGE		
VI TRL	6–8	6–8

AGROTECH

	TEMPERATURE SENSOR	PHOSPHOR TEMPERATURE SENSOR
SENSOR TYPE	Thermal	Thermal
MEASURAND	Electric current, electromagnetic light spectrum, transmitted light, temperature	Electromagnetic light spectrum, temperature
CONSTRUCTION PRINCIPLE	Thread	Thread
GEOMETRY	Linear	Linear, phosphorus diameter a few 100s of μm
MATERIAL	Metals, electrically conductive polymers, glass fibers	Doped phosphorus (Gd_2O_2S and La_2O_2S)
Procedural principle	Design of thread-shaped sensors for the investigation of thermal loads based on low-melting metal wires, which change their electrical properties under thermal load. [4] Temperature determination by measuring the change of the refraction coefficient of the light-guide sheath under temperature change, which leads to a corresponding transmission difference. [9]	Temperature determination with evaluation of the temperature-dependent luminescence of a doped phosphor at the end of an optical glass fiber to generate a luminescence, the phosphor is excited by UV light via a (multimode) fiber and the fiber guided over the same fiber is spectrally decomposed and detected. The intensity ratio of two lines is used to determine the temperature. [1]
Schematic sketch	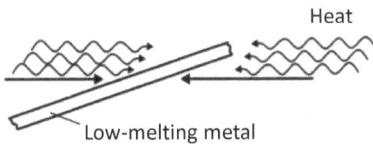 Heat Low-melting metal	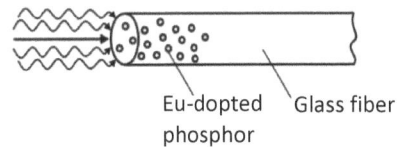 Eu-dopted phosphor Glass fiber
Known/possible field of application	Temperature monitoring of textile structures.	Temperature monitoring of textile structures.
Possible sensor variants	Use of threads of electrically conductive polymers or electrically conductive coated polymers. Temperature sensors based on the principle of absorption edge displacement, using filter glasses instead of semiconductor elements.	As an alternative to the intensity ratio, the temperature-dependent phase shift between luminescent light and excitation light can be determined with periodic excitation. The measuring range of this variant is between -30–150 °C with an accuracy of 0.04 °C. Using a small, inexpensive luminescent GaxAl1-x- As crystal as a sensor, a temperature range between 0 and 200 °C can be measured with an accuracy of 1 °C (resolution 0.1°C).
Opportunities and challenges	+ High reproducibility + Short response time + High accuracy + Due to unfavorable properties of the metals there is a low tendency for thread or surface production	+ Cost-effective + Small installation space
MATERIAL PROPERTIES		
ENERGY SUPPLY	None	None
RESOLUTION		
MATERIAL		0.1 °C
MATERIAL PROPERTIES	50–250 °C	-50–+250 °C
TRL	6–8	6–8

1	SENSOR TYPE	Thermal
2	MEASURAND	Electromagnetic light spectrum, temperature
3	CONSTRUCTION PRINCIPLE	Fiber processed into fabric
4	GEOMETRY	Flat, length of one optical waveguide up to 20 m
5	MATERIAL	Fiber made of glass, sheet material, e.g., geotextile

a) Procedural principle

Textile temperature-measuring mat with meandering optical waveguide for checking and monitoring the insulation of cladding pipe sections. The temperature is measured via a fiber-optic recording of the temperature-dependent anti-Stokes line in the optical waveguide. The temperature can be measured either continuously or sequentially by evaluating the scattered light pulses depending on the run time. From the registered temperature curve, the effectiveness of the insulation can be concluded. [11]

b) Schematic sketch

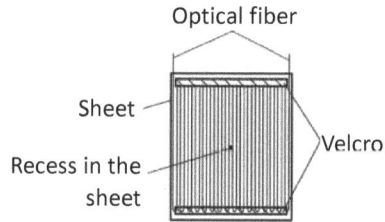

c) Known/possible field of application

Control and monitoring of the insulation of pipe sections.

d) Possible sensor variants

Replacement of the fiber-optic cable by flat distributed single sensors.

e) Opportunities and challenges

+ Simultaneous temperature measurement of several locations by means of a light pulse and the dependence of the temperature on the propagation time of the light
+ Temperature measurement already possible during the manufacture of the pipe insulation
+ Cost-effective method, since one fiber-optic cable is sufficient for temperature measurement in principle
+ Location-dependent measurement enables local weak points in the pipe insulation to be detected

I	MATERIAL PROPERTIES	Fabrics not subject to tensile, compressive and tear loads
II	ENERGY SUPPLY	Electric current
III	RESOLUTION	
IV	SENSITIVITY	0.1 K
V	MEASUREMENT RANGE	100–750 K
VI	TRL	9

	AGROTECH	
	FIBER-OPTIC DISPLACEMENT TRANSDUCER	**ACTIVE FIBER-OPTIC SENSOR**
SENSOR TYPE	Mechanical	Chemical
MEASURAND	Path, route	Visual assessment
CONSTRUCTION PRINCIPLE	Fiber-optic conductor	Fiber
GEOMETRY	Linear	Linear
MATERIAL		
Procedural principle	Measurement of paths on the basis of various principles. In particular, fiber-optic measurements of a large number of physical quantities that can be converted into paths by test specimens. [1]	Measurement of the distance between sensor and fluid environment, the concentration of chemicals in the fluid environment, the pH value of aqueous solutions, and the partial pressures of a gas by evaluating the light transmitted via the fiber-optic laser if this changes characteristically as a reaction between the sensor reagent and the surrounding environment. [12]
Schematic sketch		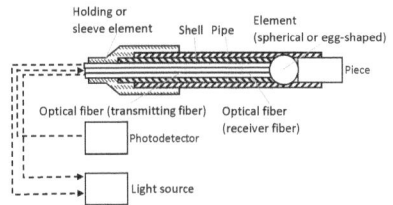
Known/possible field of application	Measurement technology, from displacement measurement, angle, pressure or acceleration can also be measured, depending on the arrangement.	Control of chemical processes in nuclear and industrial areas, underground nuclear waste in the environment, in medical and biological analysis, as well as in the agri-food industry; medical applications; biochemical applications; use in the food industry.
Possible sensor variants		Fiber-optically active sensor. [13]
Opportunities and challenges		+ Long service life + Simple sterilization + High stability − Limited pH measuring range − Limited reproducibility of the reaction between optical fibers and the immobilized reagent
MATERIAL PROPERTIES		Bulky sensor material
ENERGY SUPPLY		Light
RESOLUTION		
MATERIAL		
MATERIAL PROPERTIES	10^{-10}–1 m	
TRL	9	9

	AGROTECH	
	SOUND SENSOR (HYDROPHONE)	**RAPID-SHRINK FIBER**
1 | SENSOR TYPE	Mechanical	Chemical
2 | MEASURAND	Electric current	Visual assessment
3 | CONSTRUCTION PRINCIPLE	Fiber-optic conductor	Warp knit, weave
4 | GEOMETRY	Linear	Linear, planar
5 | MATERIAL	Quartz glass	Elastomer (polyutherane, rubber)
a) Procedural principle	Fiber-optic hydrophone (Mach–Zehnder interferometer) for highly sensitive detection of pressure differences between measuring and reference fibers. By modulating the refractive index of the measuring fiber, the sound pressure changes the phase length of the passing light and thus the interference signal, which is detected by two photodiodes and fed to the amplifier via a high-pass filter. The signal behind the low pass is used to stabilize the operating point of the interferometer against slow fluctuations, e.g., due to temperature changes. [1]	A polymer fiber which shrinks rapidly at ordinary temperature and in contact with water, but retains the fiber shape (impact strength), has high absorbency, and has performance characteristics such as rubber elasticity. [14]
b) Schematic sketch	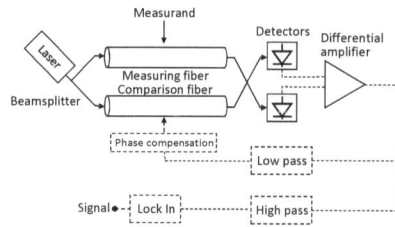	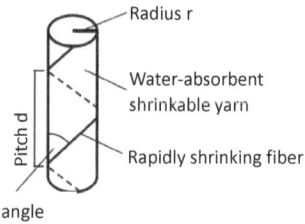
c) Known/possible field of application	Metrology.	Disposable diapers; fastening tapes; cloths as cover for dampening units in offset printers; cords or cylinders for plant cultivation; cords and nets for the food industry; bank reinforcements.
d) Possible sensor variants	Due to the flexibility of the quartz glass fibers, sensors with directional characteristics can be manufactured.	A water-absorbing, shrinkable yarn produced by blending or by blending spinning the rapidly shrinking fiber and a fiber that shrinks slower than said fiber upon the absorption of water. A water-absorbing shrinkable material which consists of a water-absorbing shrinkable fibrous web and a water-absorbing shrinkable yarn that absorbs water at a higher rate and to a greater extent than the fibrous web, with the water-absorbing shrinkable yarn containing the rapidly shrinking fiber.
e) Opportunities and challenges		
I MATERIAL PROPERTIES		At 20 °C, maximum percentage shrinkage >30%
II ENERGY SUPPLY	Laser light	
III RESOLUTION		0–10 s
IV SENSITIVITY		
V MEASUREMENT RANGE		At 20 °C, shrinkage stress = 0.351–1.755 kg/m² (30–150 mg/den)
VI TRL	9	9

BUILDTECH		
	MOISTURE- AND CHEMICAL-SENSITIVE SENSOR THREAD	CARBON-FILLED CELLULOSE PHASE
SENSOR TYPE	Mechanical, chemical	Mechanical
MEASURAND	Visual assessment	Electric current
CONSTRUCTION PRINCIPLE	n/a	Fiber, filament, film
GEOMETRY	Linear	Linear, planar
MATERIAL	n/a	Polymer
Procedural principle	Permanent identification of harmful environmental influences through the use of threads which change their shape, color or volume while absorbing liquids. The core yarn must be UV-resistant and clearly distinguished in color from the load-bearing tape. For the sheath fibers of the yarn, a material must be selected which is changed in shape, color or structure by UV radiation. [4]	Carbon-filled cellulose fiber. Detection of liquids or vapors via electrically conductive filaments from dry-wet spun cellulose dotted with charge carriers (graphite, carbon black, pigments with semiconducting layers, metallic fibers or carbon fibers) whose conductivity changes under tension/pressure or with increasing moisture content. [8]
Schematic sketch		
Known/possible field of application	A friction-spun sensor thread represents a combination of an abrasion sensor and a UV sensor.	Detection of liquids or vapors.
Possible sensor variants	Decrease in abrasion resistance with increasing exposure to UV radiation.	Mechanically stable even at high temperatures, sometimes fire-retardant.
Opportunities and challenges		− Increasing carbon-black content reduces substance strength, ductility and toughness − Doping with carbon black influences the material viscosity to such an extent that stable thread formation is not possible at normal spinning speeds − If the doping with soot is too high, the electrical resistance increases disproportionately
MATERIAL PROPERTIES		
ENERGY SUPPLY	None	Electric current
RESOLUTION		
MATERIAL		
MATERIAL PROPERTIES		
TRL	9	6–8

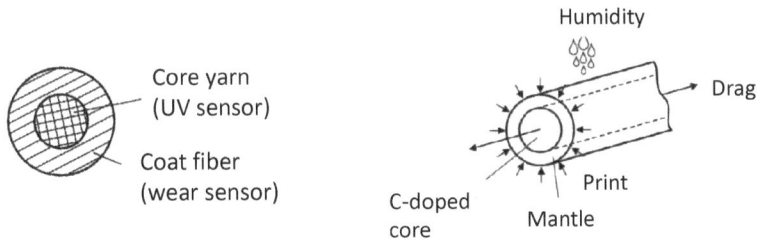

Schematic sketch (Moisture- and chemical-sensitive sensor thread):
Core yarn (UV sensor)
Coat fiber (wear sensor)

Schematic sketch (Carbon-filled cellulose phase):
Humidity
Drag
C-doped core
Print
Mantle

BUILDTECH

	INTEGRATED OPTICAL RESONATOR	TEMPERATURE SENSOR
1 | SENSOR TYPE	Thermal	Thermal
2 | MEASURAND	Electromagnetic light spectrum	Electric current, electromagnetic light spectrum, transmitted light, temperature
3 | CONSTRUCTION PRINCIPLE		Thread
4 | GEOMETRY	Linear	Linear
5 | MATERIAL	LiNbO3	Metals, electrically conductive polymers, glass fibe
a) Procedural principle	The temperature changed by means of optical resonators integrated in LiNbO3 with a periodic characteristic curve. To be able to record the number of orders passed as a function of the direction of the phase (or temperature) change requires two signals phase-shifted by 90°. It is advantageous to use the output signals to arrive at an evaluation, which counts in each case with the zero crossing, and thus an independence from slow fluctuations of the light intensity is obtained. The phase modulation required for differentiation is achieved by frequency modulation of the laser light or by electro-optical modulation of the optical path length of the resonator. [1]	Design of thread-shaped sensors for the investigation of thermal loads based on low-meltin metal wires, which change their electrical propertie under thermal load. [4] Temperature determination by measuring the change of the refraction coefficient of the light guic sheath under temperature change, which leads to a corresponding transmission difference. [9]
b) Schematic sketch	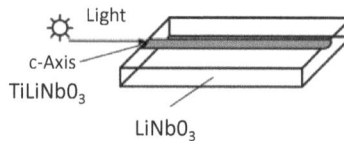 Light c-Axis TiLiNbO$_3$ LiNbO$_3$	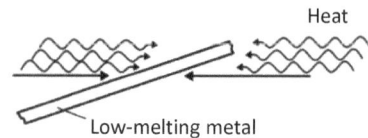 Heat Low-melting metal
c) Known/possible field of application	Temperature monitoring of textile structures.	Temperature monitoring of textile structures.
d) Possible sensor variants	The sensitivity of the temperature sensor can be determined in wide ranges by the length of the component and the wavelength of the light.	Use of threads of electrically conductive polymers or electrically conductive coated polymers. Temperature sensors based on the principle of absorption edge displacement, using filter glasses instead of semiconductor elements.
e) Opportunities and challenges	+ Simple measuring system with high accuracy when supplying the resonator sensor element via a polarization-maintaining monomode fiber + Measurement of smallest temperature changes possible due to the strong temperature dependence of the refractive index − Measurement of absolute temperatures not possible	+ High reproducibility + Short response time + High accuracy + Due to unfavourable properties of the metals low tendency for thread or surface production
I MATERIAL PROPERTIES		
II ENERGY SUPPLY	None	None
III RESOLUTION		
IV MATERIAL	Sensitivity of 35 impulses/K, resolution of 29 impulses/K	
V MATERIAL PROPERTIES		50–250 °C
VI TRL	6–8	6–8

BUILDTECH

PHOSPHOR TEMPERATURE SENSOR

SENSOR TYPE	Thermal
MEASURAND	Electromagnetic light spectrum, temperature
CONSTRUCTION PRINCIPLE	Thread
GEOMETRY	Linear, phosphorus diameter a few 100s of µm
MATERIAL	Doped phosphorus (Gd_2O_2S and La_2O_2S)
Procedural principle	Temperature determination with evaluation of the temperature-dependent luminescence of a doped phosphor at the end of an optical glass fiber to generate a luminescence, the phosphor is excited by UV light via a (multimode) fiber and the fiber guided over the same fiber is spectrally decomposed and detected. The intensity ratio of two lines determines the temperature. [1]

Schematic sketch

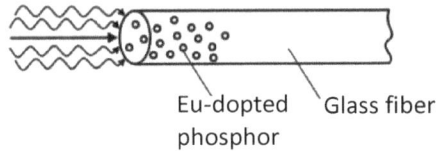

Eu-dopted phosphor   Glass fiber

Known/possible field of application	Temperature monitoring of textile structures.
Possible sensor variants	As an alternative to the intensity ratio, the temperature-dependent phase shift between luminescent light and excitation light can be determined with periodic excitation. The measuring range of this variant is between -30–150 °C with an accuracy of 0.04 °C. Using a small, inexpensive and luminescent GaxAl1-x- As crystal as a sensor, a temperature range between 0 and 200 °C can be measured with an accuracy of 1 °C (resolution 0.1 °C).
Opportunities and challenges	+ Cost-effective + Small installation space

MATERIAL PROPERTIES	
ENERGY SUPPLY	None
RESOLUTION	
SENSITIVITY	0.1 °C
MEASUREMENT RANGE	-50–+250 °C
TRL	6–8

	FIBER-OPTIC DISPLACEMENT TRANSDUCER	ALARM WALLPAPER
1 | SENSOR TYPE	Mechanical	Mechanical
2 | MEASURAND	Path, route	Electric current
3 | CONSTRUCTION PRINCIPLE	Fiber-optic conductor	Fiber fleece
4 | GEOMETRY	Linear	Planar
5 | MATERIAL		Fiber fleece: plastic; conductor paths: electrically conductive metals
a) Procedural principle	Measurement of paths on the basis of various principles. Measurement of paths. In particular, fiber optic measurements of a large number of physical quantities that can be converted into paths by test specimens. [1]	Surface monitoring system with coating of plastic fiber fleece coated with electrically conductive, metal-free conductive tracks which trigger an alarm in case of damage. [15]
b) Schematic sketch		
c) Known/possible field of application	Measurement technology, from displacement measurement, angle, pressure or acceleration can also be measured, depending on the arrangement.	Alarm in case of damage to surfaces.
d) Possible sensor variants		Simple retrofitting possible.
e) Opportunities and challenges		+ Self-calibration function + Device not detectable via instruments + Side effects (noises, vibrations and temperature fluctuations) are not recorded + Modular system structure possible + Roll material for use in all cases of need − Impairment by nails, screws or dowels in the wall
I MATERIAL PROPERTIES		
II ENERGY SUPPLY		Electric current
III RESOLUTION		
IV SENSITIVITY		
V MEASUREMENT RANGE	10^{-10}–1 m	
VI TRL	9	6–8

BUILDTECH

	LUMINOUS SIGNAL FILAMENT	LAMELLA
SENSOR TYPE	Chemical	Mechanical
MEASURAND	Visual assessment	Electromagnetic light spectrum
CONSTRUCTION PRINCIPLE	Friction-spun yarn	Composite material; weave, warp knit, weft knit, netting, scrim
GEOMETRY	Linear	Flat, lamella 400 mm x 200 mm
MATERIAL	Polypropylene core, polypropylene or polyester jacket	Optical fiber: polymer or glass; carrier textile: glass, carbon, aramid, or basalt scrim
Procedural principle	Friction-spun yarn or wrap-around yarn with a light-intensive signal thread (with color and light effects) visibly integrated into the core from the outside for the detection of a wear condition. The signal thread is covered by a cover sensitive to environmental influences (abrasion, UV radiation, chemicals), which is why this is visually recognizable after exceeding a limit load adjustable via the resistance of the cover. [4]	Embroidered arrangement of fiber-optic high-performance fibers with integrated fiber-Bragg-gratings on a lamella for detection of temperature changes, elongations, compressions and oscillations in supporting structures. Guiding the fiber-optic sensor in the direction of the lines of force for the detection of tensile, compressive and shear forces and also transversely to the lines of force for temperature compensation. Solidification of the textile structure by means of a resin system and construction of the composite material from one or more textile layers. [16]
Schematic sketch	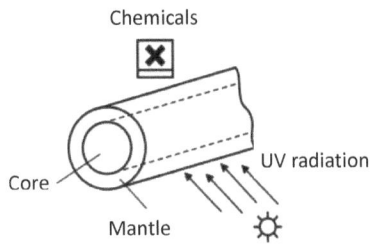 Chemicals / Core / Mantle / UV radiation	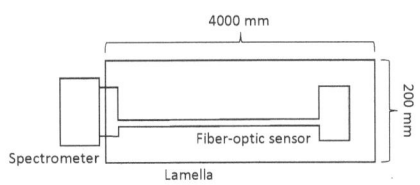 4000 mm / 200 mm / Spectrometer / Fiber-optic sensor / Lamella
Known/possible field of application	Inspection of the wear condition of belts and ropes by means of camera observation.	Reinforcement and monitoring of concrete and wooden structures. Critical deflection of structural elements. Critical crack formation. Evidence of functionality, reliability and safety evidence of remaining useful life.
Possible sensor variants		Incorporation of fiber Bragg gratings before or after textile processing.
Opportunities and challenges		+ Fast and reliable application for building refurbishments + No temperature dependence
MATERIAL PROPERTIES	Sensitive sheath, fluorescent core	
ENERGY SUPPLY		
RESOLUTION		
MATERIAL		
MATERIAL PROPERTIES		
TRL	9	9

	BUILDTECH	
	TEMPERATURE-CONTROLLED RADIATION TRANSMISSION	**PYROMETERS**
1 | SENSOR TYPE	Mechanical	Thermal
2 | MEASURAND	Electric current	Electromagnetic light spectrum, temperature
3 | CONSTRUCTION PRINCIPLE	Fleece	Fiber-optic conductor
4 | GEOMETRY	Flat, fiber diameter 0.01 to 10 mm	Linear
5 | MATERIAL	Thermotropic polymer blends	Sapphire glass, quartz glass
a) Procedural principle	A polymer-based material having temperature-controlled radiation transmission which is present within core/sheath fibers in a core. A transparent shell surrounds the core of thermotropic polymer mixture, which becomes turbid beyond the so-called lower critical demixing temperature (LCST) due to changing radiation emission. This turbidity effect occurs due to a structural change in the polymer system, in which the components with different refractive indices separate due to temperature change. A variation of the relative contents of the individual comonomers causes turbidity at different temperatures. [17]	Fiber-optic measurement method that determines the temperature by analyzing the cavity radiation of a black body. The radiation spectrum of the black body shifts according to Planck's law of radiation depending on temperature. [1]
b) Schematic sketch	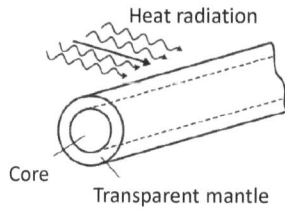 Heat radiation / Core / Transparent mantle	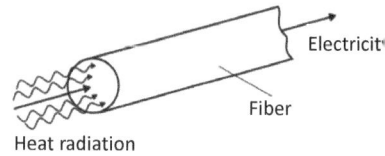 Electricity / Fiber / Heat radiation
c) Known/possible field of application	Temperature-dependent control of radiation transmission on buildings (awnings, roller blinds, venetian blinds), technical equipment, in the clothing industry and for decorative purposes.	Non-contact temperature measurement.
d) Possible sensor variants	Incorporation of a non-thermotropic but mechanically highly resilient material into the polymer core.	Very small heat capacity allows measurement of rapid temperature changes.
e) Opportunities and challenges	+ Advantage of core-shell structure when using aids with low compatibility to thermotropic core material − Expensive production − Bonding of polymers only possible at high application temperatures − Limited possibility of reversible structural change − Low mechanical-load capacity	+ Measurement of very high temperatures possible
I MATERIAL PROPERTIES	Relative proportion of comonomers between 0.1 and 50 mol%	
II ENERGY SUPPLY	Electromagnetic	Electric current
III RESOLUTION		
IV SENSITIVITY		Measurement accuracy of 0.05%
V MEASUREMENT RANGE		Up to about 2000 °C
VI TRL	9	9

BUILDTECH

	HYBRID ROPE	GLASS LIGHT CONDUCTOR
SENSOR TYPE	Mechanical	Mechanical
MEASURAND	Electromagnetic light spectrum	Electromagnetic light spectrum
CONSTRUCTION PRINCIPLE	Fiber	Fiber-optic conductor
GEOMETRY	Linear	Linear
MATERIAL	Protective layer: cotton; light-guide sheath and core: fluoride glass	Glass
Procedural principle	Investigation of the wear condition of the load-bearing rope by evaluating the ratio of the refractive index between rope core and sheath. The rope is composed of several modules of different properties, with at least one module A having the primary load-bearing function and the secondary driving function and one module B having the primary driving function and the secondary load-bearing function. By inserting conductive elements into the non-conductive modules and sensors, rope elongation can be measured by determining the position of a counterweight. [18]	Determination of the wear condition in ropes and belts by evaluating the light transmitted in the optical waveguide. Single-mode and multimode fibers transport light of a certain wavelength or light of different wavelengths depending on their wear condition. [4]
Schematic sketch	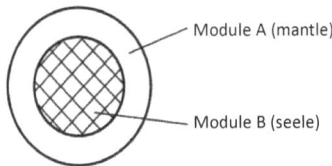 Module A (mantle); Module B (seele)	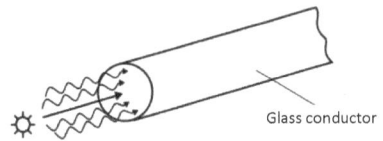 Glass conductor
Known/possible field of application	Structural health monitoring of ropes.	Use as control tearing thread in ropes and tapes with low elongation at break.
Possible sensor variants	Targeted analysis of individual, specific substances with desired concentration content.	Problems in further textile processing due to buckling sensitivity.
Opportunities and challenges	+ High response sensitivity, even to individual media only + High mechanical strength + Low manufacturing and general cost	
MATERIAL PROPERTIES	Light-guide sheath with lower refractive index than the conductor core Light-guide sheath made of fluoride glass with lower hydrolytic resistance	Elongation at break up to 5%
ENERGY SUPPLY		
RESOLUTION		
SENSITIVITY		
MEASUREMENT RANGE		
TRL	6–8	6–8

	SENSOR THREAD WITH COLOR AND LIGHT EFFECTS	WEAR SENSOR
1 | SENSOR TYPE	Mechanical, chemical	Mechanical
2 | MEASURAND	Visual assessment	Visual assessment
3 | CONSTRUCTION PRINCIPLE		Thread
4 | GEOMETRY	Linear	Linear
5 | MATERIAL		
a) Procedural principle	Permanent signals of loading and wear of the material are visualized without the supply of auxiliary energy by generating the following effects: decomposition of the sensor thread, change in color, shape or volume (swelling, shrinkage, crimping, bending), turbidity or change in mechanical properties (e.g., embrittlement by UV radiation). The preferred design form is the core-sheath structure of friction-spun wrapping yarns, in which after the destruction of the sensor material arranged in the sheath a luminous signal thread arranged in the core becomes visible. [4]	Visual assessment of wear by binding colored threads under the fabric surface of tapes and ropes. If wear occurs, the colored threads become visible on the surface. [4]
b) Schematic sketch	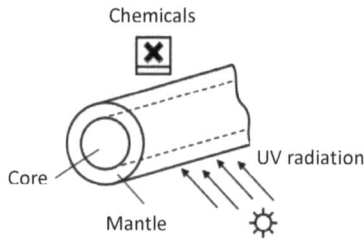 Chemicals ⊠ Core — Mantle — UV radiation	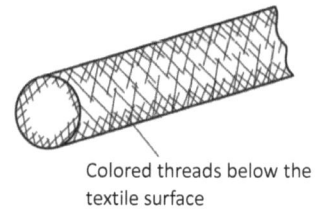 Colored threads below the textile surface
c) Known/possible field of application	Structural health monitoring of ropes.	Structural health monitoring of ropes.
d) Possible sensor variants		
e) Opportunities and challenges		
I MATERIAL PROPERTIES		
II ENERGY SUPPLY		
III RESOLUTION		
IV SENSITIVITY		
V MEASUREMENT RANGE		
VI TRL	6–8	9

BUILDTECH

	HEAT-SENSITIVE SENSOR FILAMENT	UV SENSOR FIBER
SENSOR TYPE	Thermal	Chemical
MEASURAND	Visual assessment	Visual assessment
CONSTRUCTION PRINCIPLE	Thread	Thread
GEOMETRY	Linear	Linear
MATERIAL	Viscose, polypropylene, polyester, polyamide	
Procedural principle	Permanent proof of heat exposure to load-bearing ropes and tapes at impermissibly high temperatures through the use of filamentary heat sensors. The thermochromic sensor material changes its color, shape and stiffness reversibly or irreversibly under the influence of heat. The effects mentioned appear as color change, thread crimping or activation of hot melt yarns. The latter consist of copolyamide or copolyester as a whole or as a combination yarn in parts and show a strong thread crimp depending on the temperature and the different melting ranges. [4]	Permanent signaling of the reaching or exceeding of a maximum permissible limit for the effect of UV radiation on load-bearing belts and ropes by accumulation sensors. In contrast to photochromic materials, which only record the instantaneous radiation intensity, accumulation sensors visualize the total measure of the radiation effect. [4]
Schematic sketch		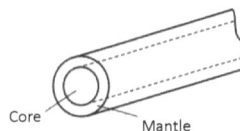 Core Mantle
Known/possible field of application	Inspection and monitoring of ropes and belts at deflection points such as eyelets or pulleys, where they are exposed to increased mechanical and thermal stress due to friction, as well as at points subject to other environmental influences with a high temperature effect.	
Possible sensor variants	Effect visualization through shape change is a cost-effective alternative to visualization through color change.	Core-sheath structures in the form of friction and wrapping yarns, which consist of a UV-sensitive sheath (sensor thread) and a luminous signal thread in the core analogous to the abrasion-sensitive sensor threads. Twisted yarns consisting of two or more threads with almost identical (colorimetrically adjusted) hues but different light fastness, which change their appearance from self-colored to multicolored after UV irradiation by bleaching of the threads with lower light fastness.
Opportunities and challenges		+ Semi-quantitative determination of the radiation dose using the reference filament + The elimination of twine production in one additional operation means that the titre of the individual yarns can also be adjusted to the yarns used in the product + Both sensor thread and reference thread can be processed individually in adjacent positions in the tape fabric or braid, provided they are suitable for the weave
MATERIAL PROPERTIES		
ENERGY SUPPLY	Heat	Electromagnetic radiation
RESOLUTION		
SENSITIVITY		
MEASUREMENT RANGE		
TRL	9	9

	STRAIN SENSOR	CONTROL TEAR STRIP
1 | SENSOR TYPE	Mechanical	Mechanical
2 | MEASURAND	Electric current	Visual assessment, electric current
3 | CONSTRUCTION PRINCIPLE	Thread	Thread
4 | GEOMETRY	Linear, diameter 0.5–2.5 mm	Linear
5 | MATERIAL	Kevlar, carbon-black-filled silicone rubber	Polyester, (silver-plated) polyamide, metallic fine wires, cellulose fiber filled with carbon, glass
a) Procedural principle	Measuring arrangement for determining the strain state in ropes. Based on the location of metal balls incorporated at defined distances by electromagnetic means, the strain results from the distance and the traversing speed of the balls, since these variables are associated with a change in the specific electrical parameters. [4]	Permanent indication of a one-time load overrun of a belt due to the failure of a control tear thread at a defined elongation value which is significantly below the elongation at break of the belt. [4]
b) Schematic sketch		
c) Known/possible field of application	Detection of individual wire breaks in steel ropes, e.g., in Kevlar elevator ropes. Use for in situ monitoring and determination of load cycles.	Structural health monitoring of ropes.
d) Possible sensor variants	Measurement of strains and strain peaks on the basis of a reproducible dependence on strain and electrical resistance, also while maintaining the strain state by plastic deformation. [4]	Non-conductive control tear thread: Consisting of textile materials such as polyester or polyamide, whose geometric integration into the textile load handling attachment is decisive for the elongation of the overall system at which failure occurs. Detections of a few percent can be realized by means of control yarns with non-typical textile elongations such as carbon fiber, glass fiber or Twaron aramid filament yarn.
e) Opportunities and challenges	+ For protection against overloading, it is not necessary for the sensor thread to fail. Exceeding a defined strain state is sufficient for the output of an alarm signal + By also detecting strain peaks, strain sensors open up a wide range of applications, from crack sensors to sensors for detecting strain peaks − Process cannot be applied to man-made fiber tapes and ropes	+ Silver-coated polymer thread: unsuitable as electrically conductive control tearing thread, since elongations at break cannot be reproduced or the parallel position of the untwisted filaments results in only individual filaments tearing in case of failure and the applied tension remaining constant − Metallic fine wire: very sensitive to breakage, otherwise excessively high elongation at break compared to load-bearing agent − Cellulose fiber with carbon filling: lower, moisture-dependent conductivity than silver-plated polyamide yarns or fine wires. − Optically conductive control thread: buckling sensitivity, critical mechanical behavior.
I MATERIAL PROPERTIES	Hardness: 50 ± 5 Shore A; density: 1.13 g/cm³; tear strength: ≥3.5 N/mm²; elongation at break: ≥200%; specific volume resistance: ≤12 Ω cm	
II ENERGY SUPPLY	Electric current	Electric current
III RESOLUTION		
IV SENSITIVITY		
V MEASUREMENT RANGE		
VI TRL	9	9

Schematic sketch (STRAIN SENSOR): Distance d, Metal sphere, Textiles carrier band, Speed v, Induction sensor

Schematic sketch (CONTROL TEAR STRIP): Tension, Weight breakage

BUILDTECH

	SENSOR FOR DETERMINING MANTLE SLIPPAGE	FRICTION SPUN ABRASION SENSOR THREAD
SENSOR TYPE	Mechanical	Mechanical
MEASURAND	Electric current	Visual assessment
CONSTRUCTION PRINCIPLE	Thread, weave	Thread
GEOMETRY	Linear, planar	Linear
MATERIAL		Polypropylene, polyethylene terephthalate
Procedural principle	Sensor with optical signal output in the event of critical wear or damage to the outer sheath of a load-bearing rope or tape. The sensor thread has a core-sheath structure, the signal-colored core being sheathed with thermoplastic staple fiber. This has the color of the load-bearing textile and is integrated into its outer shell in such a way that it is exposed to abrasion during use. [4]	Sensor with optical signal output in the event of critical wear or damage to the outer sheath of a load-bearing rope or tape. The sensor thread has a core-sheath structure, the signal-colored core being sheathed with thermoplastic staple fiber. This has the color of the load-bearing textile and is integrated into its outer shell in such a way that it is exposed to abrasion during use. [4]
Schematic sketch	Core / Mantle / Electricity / Relative movement between mantle and core	Signal-colored core / Mantle of staple fibers in color of the load-bearing band
Known/possible field of application	Control and monitoring when guiding a rope with small deflection radii, since there can be relative movements between the core and sheath in the form of sheath slippage or compression and the introduction of forces into the textile sheath, which is designed as a non-load-bearing element, leads to impermissible wear on the rope.	Structural health monitoring of ropes.
Possible sensor variants	Arrangement of the conductor loops in dimensions that correspond to the pole configuration of a planar permanent magnet. Conductor loops as an execution of adjacent meshes through which a sectionally magnetized longitudinal structure can be moved. By matching conductor loops and permanent magnets, the induction voltages of all conductor loops add up.	Variation of the ratio of core to shell diameter. Core yarn made of polyethylen (PET), sheath yarn made of polypropylene (PP); core thread not signal-colored, but made of fluorescent material for UV detection. Variation of core and sheath strength. Core yarn made of PP, sheath yarn made of PET.
Opportunities and challenges	+ Accelerations due to relative movements of core and mantle are shown in a manner directly proportional to the magnitude of the stress induced in the meshes of the mantle + The magnetized threads can be oriented along the expected displacement and anchored to the core	+ The use of a fluorescing signal thread in the thread core enables an automated, visual inspection of the wear condition by means of camera technology, even on soiled or very colorful load-bearing textiles + With increasing sheath fineness, there is a significant increase in bearable double chafing
MATERIAL PROPERTIES		
ENERGY SUPPLY	None	None
RESOLUTION		
SENSITIVITY		
MEASUREMENT RANGE		
TRL	9	6–8

PAGE 96	FIBER-OPTIC MICRO STRAIN SENSOR	ADAPTIVE FIBER COMPOSITES (ADAPTRONICS)
1 | SENSOR TYPE	Mechanical	Mechanical
2 | MEASURAND	Electric current	Vibration, deformation
3 | CONSTRUCTION PRINCIPLE	Composite material of filament yarns with embedded sensors	
4 | GEOMETRY	Planar	
5 | MATERIAL		
a) Procedural principle	Strain measurements of individual filaments and in a surrounding concrete matrix, in combination with Fabry–Pérot fiber interferometer sensors to measure dynamic events (acoustic emission, crack formation, etc.) in the matrix. [19]	Active vibration suppression by piezoelectric films and fibers which self-adjust to changing componer vibrations and deformations by integrated sensors, and also initiate counter-signals via actuators into the textile structure. [20]
b) Schematic sketch	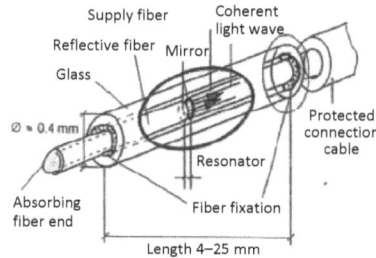	
c) Known/possible field of application	Structural health monitoring of ropes.	
d) Possible sensor variants	Use of fiber-Bragg-grating sensors not possible due to their stiffness.	Lightweight construction possible; high stiffness ar high-strength fiber composites.
e) Opportunities and challenges	− Use of strain gauges not possible due to very small dimensions − The bond of the sensors with the filaments and the matrix must be ensured − Embedded sensors must not impede the deformation of filaments and matrix	− Fundamentally low mechanical resistance to noise and vibration
I MATERIAL PROPERTIES		
II ENERGY SUPPLY	Electric current	None
III RESOLUTION		
IV SENSITIVITY	0.5 µm/m	
V MEASUREMENT RANGE		
VI TRL	<6	9

Schematic sketch labels: Supply fiber, Coherent light wave, Reflective fiber, Mirror, Glass, Ø ≈ 0.4 mm, Protected connection cable, Resonator, Absorbing fiber end, Fiber fixation, Length 4–25 mm

BUILDTECH

	STRAIN/PRESSURE SENSOR	INTELLIGENT MEMBRANE
SENSOR TYPE	Mechanical	Mechanical
MEASURAND	Electric current	Electromagnetic light spectrum, electric current, noise level
CONSTRUCTION PRINCIPLE	Weft knit, warp knit	Fiber bundle in woven or knitted structure
GEOMETRY	Planar	Areal, change of shape up to 8 times its size
MATERIAL	Stainless steel	Nickel titanium alloy
Procedural principle	Spacer weft-knit made of electrically conductive stainless-steel-fiber yarns for detecting the position of the contact and the size of the contacting surface when the specific electrical resistance of the electrically conductive conductor paths changes as a result of elongation or pressure. [21]	A membrane with built-in sensors which reacts to stimuli such as light, contact, noise or environmental movements in a mobile manner via muscle wires made of Ni-Ti alloy developing different temperatures at certain currents and passing through different movements. [22]
Schematic sketch		

Knitted surface
Polfiber

Known/possible field of application	Flat pressure load in buildings.	
Possible sensor variants	Spacer warp-knit can also be used instead of spacer weft-knit.	Use in any size possible.
Opportunities and challenges	– Spacer warp-knit has a hysteretic force behavior and is therefore less suitable as a pressure sensor	– Very expensive materials
MATERIAL PROPERTIES		
ENERGY SUPPLY	Electric current	Electric current
RESOLUTION		
SENSITIVITY		
MEASUREMENT RANGE		
TRL	6–8	<6

	WEAR SENSOR 2	STRAIN SENSOR 2
1 | SENSOR TYPE	Mechanical	Mechanical
2 | MEASURAND	Electric current	Electric current
3 | CONSTRUCTION PRINCIPLE	Thread	Thread
4 | GEOMETRY	Linear	Linear, diameter ø = 0.5–2.5 mm
5 | MATERIAL	Aramid	Carbon-black-filled silicone rubber
a) Procedural principle	Measurement of wear by a disruption sensor made of aramid. Carbon fibers integrated into a rope are contacted electronically. As the disintegration progresses, the electrical resistance increases, which serves as a characteristic value for wear.	Measurement of strains and strain peaks based on a reproducible dependence on strain and electrical resistance, even while maintaining the state of strain through plastic deformation.
b) Schematic sketch	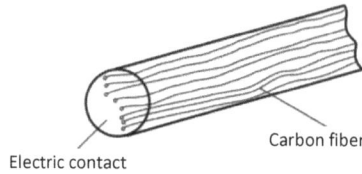 Electric contact / Carbon fiber	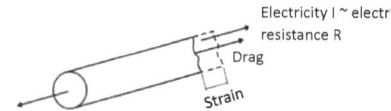 Electricity I ~ electr resistance R / Drag / Strain
c) Known/possible field of application		Use for in situ monitoring and determination of loa cycles.
d) Possible sensor variants		
e) Opportunities and challenges		+ For protection against overloading, it is not necessary for the sensor thread to fail. Exceeding a defined strain state is enough for the output of an alarm signal + By also detecting strain peaks, strain sensors open a wide range of applications, from crack sensors to sensors for detecting strain peaks
I MATERIAL PROPERTIES		Hardness: 50 ± 5 Shore A; density: 1.13 g/cm^3; tear strength: ≥3.5 N/mm^2; elongation at break: ≥200%; specific volume resistance: ≤12 Ω cm
II ENERGY SUPPLY	Electric current	Electric current
III RESOLUTION		
IV SENSITIVITY		
V MEASUREMENT RANGE		
VI TRL	6–8	9

	BUILDTECH	
	FIBER-COATED SENSORS	**INTENSITY-BASED FIBER-OPTIC SENSORS**
SENSOR TYPE	Chemical	Chemical
MEASURAND	Electromagnetic light spectrum	Electromagnetic light spectrum
CONSTRUCTION PRINCIPLE	Weave	Fiber
GEOMETRY	Planar	Linear
MATERIAL	Fiber Bragg Grating (FBG) fibers in fabric	Optically conductive fibers
Procedural principle	FBG is a distributed Bragg reflector constructed in a short segment of optical fiber that reflects certain wavelengths of light and transmits all others. This is achieved by creating a periodic variation in the refractive index of the fiber core, which generates a wavelength-specific dielectric mirror. [23]	A measurand-induced change in the optical intensity propagated by an optical fiber can be produced by different mechanisms, such as micro-bending loss, attenuation and evanescent fields. Requires lighter fibers. They usually use multi-mode large-core fibers. [24]
Schematic sketch		
Known/possible field of application	Used in seismology, pressure sensors for extremely harsh environments, and downhole sensors in oil and gas wells for the measurement of the effects of external pressure, temperature, seismic vibrations and inline flow measurement.	
Possible sensor variants	Integration of Bragg fiber as warp thread; or into a 3D woven; embedded in a conveyor belt; inserted into a groove and threaded into flat-woven fabric.	
Opportunities and challenges	+ Inline optical filter to block certain wavelengths, or as a wavelength-specific reflector	+ Advantages of this category are easy implementation, low cost, multiplexing, and the possibility to implement distributed sensors − Disadvantages include the measurements and variations in the intensity of the light source, which could lead to false readings if a reference system is not used
MATERIAL PROPERTIES		
ENERGY SUPPLY	Light	Light
RESOLUTION		
SENSITIVITY		
MEASUREMENT RANGE		
TRL	<6	6–8

	BUILDTECH	
	INTRINSIC DISTRIBUTED FIBER-OPTIC SENSORS	**FBG SENSORS**
1 | SENSOR TYPE	Chemical	Chemical
2 | MEASURAND	Electromagnetic light spectrum	Electromagnetic light spectrum
3 | CONSTRUCTION PRINCIPLE	Fiber	Fiber
4 | GEOMETRY	Linear	Linear
5 | MATERIAL	Optically conductive fibers	Optically conductive fibers
a) Procedural principle	Based on Rayleigh scattering. The light is subjected to attenuation due to this scattering, which is determined by random microscopic variations. If a narrow optical pulse is launched in the fiber, it is possible to determine the spatial variations in the fiber scattering coefficient or the attenuation by monitoring the variation of the Rayleigh backscattered signal intensity. The scattering coefficient of a location is influenced by the local fiber status. [24]	FBGs are characterized by periodic changes created by an intense interference pattern of UV energy in the index of refraction in the core of a single-mode optical fiber. The grating reflects a spectral peak based on the grating spacing; therefore, a variation in the length of the fiber due to tension or compression determines a change in the grating spacing and consequently in the wavelength of light that is reflected. By measuring the center wavelength of the reflected spectral peak, it is possible to obtain a quantitative measurement of the strain. [24]
b) Schematic sketch		
c) Known/possible field of application		
d) Possible sensor variants	Raman scattering is a phenomenon which involves the inelastic scattering of photons. The incident light pulse causes molecular vibrations in the optical fiber. In the case of optical time-domain reflectometry (OTDR), a high-input power is requested, as the Raman scattering coefficient is about three orders of magnitude lower than the Rayleigh scattering coefficient. Brillouin scattering is caused by the acoustic vibrations that occur in the optical fiber when an optical pulse is launched. OTDR in different approaches: OTDR based on Rayleigh scattering, OTDR based on Raman scattering, OTDR based on Brillouin scattering. Raman scattering is used for the development and implementation of reliable distributed temperature sensors. Rayleigh scattering is used to track and to reveal propagation effects.	The response of several FBG sensors can be measured simultaneously by placing several networks in series attached to one lead optical fiber. This is a relevant advantage with respect to traditional strain sensor measurement, which requires an acquisition system for each sensor. By using different wavelengths that are reflected, various FBG sensor signals can be identified, and therefore the space-distributed sensors are identified and distinguished. An optical switch must then be used to connect several optical fibers to the light source and the spectrometer that measures the reflected wavelengths. The direct embedding of optical fibers with FBG in the epoxy resin of fiber-reinforced polymer (FRP) materials allows exact strain measurement in the material. Therefore, epoxy resin is an effective protection for the optical fiber. Used for quasi-distributed measurement of strain.
e) Opportunities and challenges		
I MATERIAL PROPERTIES		
II ENERGY SUPPLY	Light	Light
III RESOLUTION		
IV SENSITIVITY		
V MEASUREMENT RANGE		
VI TRL	6–8	6–8

BUILDTECH

TEXTILE-BASED GONIOMETER

SENSOR TYPE	Mechanical
MEASURAND	Electric current
CONSTRUCTION PRINCIPLE	Weft knit
GEOMETRY	Planar
MATERIAL	Combination of electroconductive (Belltron® by Kanebo Ltd.) and elastic (Lycra®) yarns
Procedural principle	Knitted piezoresistive fabrics modify their electrical resistance when they are elongated or flexed. The main requirement for the application of the single-layer sensors is that the human movements must produce a strain field which can be detected in terms of resistance variation. For this reason, single-layer sensors must be integrated into adherent garments close to the human joint under investigation. [25]
Schematic sketch	

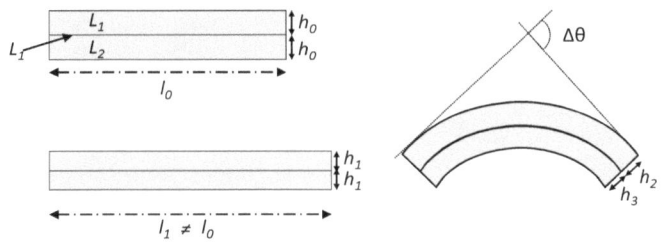

Known/possible field of application	Hand motion sensing: a kinesthetic sensing glove was developed for the ambulatory evaluation of the residual hand function and its recovery in post-stroke patients; scapular movement detection.
Possible sensor variants	Single-layer sensor or double-layer sensors.
Opportunities and challenges	– Needs to closely adhere to joint moving

MATERIAL PROPERTIES	
ENERGY SUPPLY	Electric current
RESOLUTION	
SENSITIVITY	Single layer: 6405 Ω for $\Delta\theta = 37°$
MEASUREMENT RANGE	Double layer: 5100 Ω for $\Delta\theta = 37°$
TRL	6–8

CLOTHTECH

	HUMIDITY SENSOR	MOISTURE- AND CHEMICAL-SENSITIVE SENSOR THREAD
1 | SENSOR TYPE	Chemical	Mechanical, chemical
2 | MEASURAND	Electric current	Visual assessment
3 | CONSTRUCTION PRINCIPLE	Weft knit	n/a
4 | GEOMETRY	Planar	Linear
5 | MATERIAL	Electrically conductive yarn	n/a
a) Procedural principle	Knitted fabric with a basic weft knit which contains at least one thread made of a material which changes its electrical resistance when affected by moisture. The weft knit is equipped with an integrated moisture sensor consisting of at least two electrodes arranged at a distance, which are electrically connected to each other in case of moisture. [2]	Permanent identification of harmful environmental influences through the use of threads which change their shape, color or volume while absorbing liquid. The core yarn must be UV-resistant and clearly distinguished in color from the load-bearing tape. For the sheath fibers of the yarn, a material must be selected which is changed in shape, color or structure by UV radiation. [4]
b) Schematic sketch		

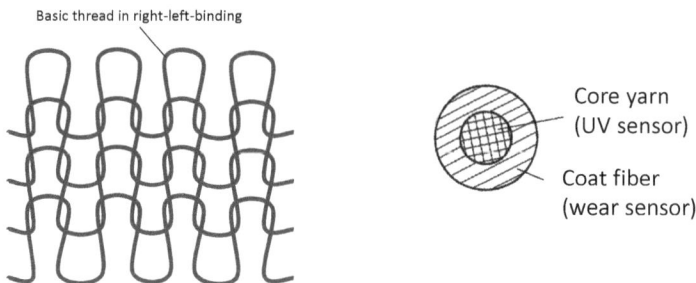

Basic thread in right-left-binding

Core yarn (UV sensor)

Coat fiber (wear sensor)

	HUMIDITY SENSOR	MOISTURE- AND CHEMICAL-SENSITIVE SENSOR THREAD
c) Known/possible field of application	Woven fabrics in which electrically wellconducting and electrically not-well-conducting threads are alternately woven with each other. Electrical connection means in the form of terminals, plug-in connection parts.	A friction-spun sensor thread represents a combination of an abrasion sensor and and a UV sensor.
d) Possible sensor variants	Electrical means of connection can be connection to the monitoring station via textile conductors. The textile behavior ensures that the joint is extremely flexible and elastic.	Decrease in abrasion resistance with increasing exposure to UV radiation.
e) Opportunities and challenges	+ Integration of the sensor directly into the garment, with no external application necessary	
I MATERIAL PROPERTIES		
II ENERGY SUPPLY	Electric current	None
III RESOLUTION		
IV MATERIAL		
V MATERIAL PROPERTIES		
VI TRL	6–8	9

CLOTHTECH

TEMPERATURE SENSOR

| SENSOR TYPE	Thermal
| MEASURAND	Electric current, electromagnetic light spectrum, transmitted light, temperature
| CONSTRUCTION PRINCIPLE	Thread
| GEOMETRY	Linear
| MATERIAL	Metals, electrically conductive polymers, glass fibers

Procedural principle

Design of thread-shaped sensors for the investigation of thermal loads based on low-melting metal wires, which change their electrical properties under thermal load. [4]

Temperature determination by measuring the change of the refraction coefficient of the light-guide sheath under temperature change, which leads to a corresponding transmission difference. [9]

Schematic sketch

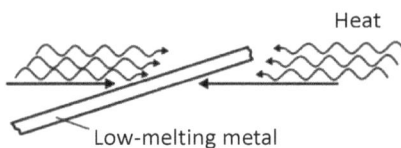

Heat

Low-melting metal

Known/possible field of application

Temperature monitoring of textile structures.

Possible sensor variants

Use of threads made of electrically conductive polymers or electrically conductive coated polymers. Temperature sensors based on the principle of absorption edge displacement using filter glasses instead of semiconductor elements.

Opportunities and challenges

+ High reproducibility
+ Short response time
+ High accuracy
+ Due to unfavourable properties of the metals low tendency for thread or surface production

MATERIAL PROPERTIES	
ENERGY SUPPLY	None
RESOLUTION	
SENSITIVITY	
MEASUREMENT RANGE	50–250 °C
TRL	6–8

PAGE 104	STRAIN/PRESSURE SENSOR	TEXTILE NETTLE CELL
1 | SENSOR TYPE	Mechanical	Thermal
2 | MEASURAND	Electric current	Temperature
3 | CONSTRUCTION PRINCIPLE	Weft knit, warp knit	Wire, integrated in support fabric
4 | GEOMETRY	Planar	Planar
5 | MATERIAL	Stainless steel	Shape-memory metal
a) Procedural principle	Spacer weft-knit made of electrically conductive stainless-steel-fiber yarns for detecting the position of the contact and the size of the contacting surface when the specific electrical resistance of the electrically conductive conductor paths changes as a result of elongation or pressure. [21]	Implementation of the textile sensor in the initial fabric by which it autarkically warns the wearer of excessive heat stress on the outside of the garment due to irritation on the inside of the fabric. Heat collectors (metal plates) pass heat onto a heat insulator (time delay element), which delivers a defined amount of heat to a rolled blunt needle made of shape-memory metal (Nitinol). With enough heat, the needle stretches through the undergarment and irritates the skin of the wearer. [26]
b) Schematic sketch	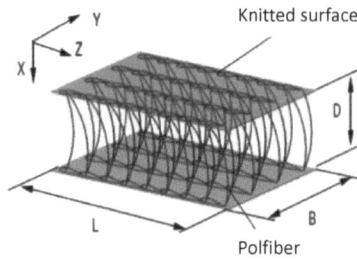	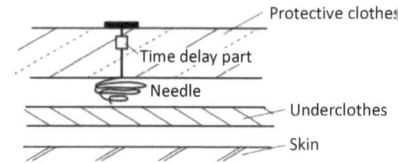
c) Known/possible field of application	Flat pressure load in buildings.	Personnel potentially exposed to high temperature
d) Possible sensor variants	Spacer warp-knit can also be used instead of spacer weft-knit.	Simple configuration of sensor sensitivity via material selection.
e) Opportunities and challenges	− Spacer warp-knit has a hysteretic force behavior and is therefore less suitable as a pressure sensor	+ Cost-effective and unproblematic made-to-measure clothing + Self-sufficient and redundant system + No susceptible cabling + Fast location and size estimation of the heat source + No warning signals need to be monitored continuously + Having few layers of clothing prevents greater heat stress
I MATERIAL PROPERTIES		
II ENERGY SUPPLY	Electric current	Heat
III RESOLUTION		
IV SENSITIVITY		
V MEASUREMENT RANGE		
VI TRL	6–8	6–8

	CLOTHTECH	
	TEXTILES WITH SPECIAL FUNCTIONS	**INDICATIVE COLOR SENSOR**
SENSOR TYPE	Chemical	Chemical
MEASURAND	Visual assessment, electric current	Visual assessment
CONSTRUCTION PRINCIPLE		Printed fabric
GEOMETRY		Punctiform and areal
MATERIAL	Hollow polymers modified with moisture-sensitive gels	Fluorescent agent
Procedural principle	Garment comprising sensory and/or actuatorically modified polymers which, in the event of a health and/or environmental hazard, change their color, geometric shape or other physical, biological or chemical properties to protect the wearer in a defined manner. [27]	A light-sensor layer, temperature sensor layer and fluorescent layer with applied writing, pattern or three-dimensional form, which change their shape and aesthetic impression when externally influenced. [28]
Schematic sketch	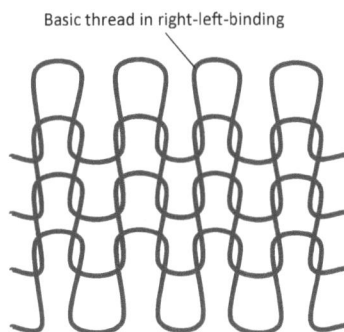Basic thread in right-left-binding	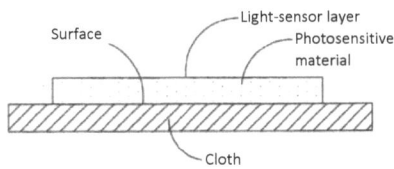Surface — Light-sensor layer — Photosensitive material — Cloth
Known/possible field of application	Monitoring of dangerous conditions.	Light-sensor layer for detection of UV radiation. Temperature sensor layer for temperature determination. Fluorescent layer for generating fluorinating light.
Possible sensor variants	Sensor element can be formed from: temperature sensors, pressure sensors, humidity sensors, pH sensors, radiation sensor.	
Opportunities and challenges		+ Optically appealing design of signal bodies
MATERIAL PROPERTIES		
ENERGY SUPPLY		Light and heat
RESOLUTION		
SENSITIVITY		
MEASUREMENT RANGE		
TRL	6–8	6–8

	CLOTHTECH	
	CLOTHING INDICATOR FOR UV RADIATION AND OZONE	**CBRN PROTECTIVE CLOTHING**
1 | SENSOR TYPE	Chemical	Mechanical, chemical, thermal
2 | MEASURAND	Visual assessment	Electric current
3 | CONSTRUCTION PRINCIPLE		Weft knit, warp knit, weave, scrim, fleece and composite fabrics
4 | GEOMETRY		Punctiform, linear or planar
5 | MATERIAL		
a) Procedural principle	Small-scale application of a variety of fashionable design forms whose color change is accompanied by influencing factors from the environment and at least semi-quantitatively correlates with the hazard potential. The measurement is carried out either by the iodine method, acetone decomposition, oxalic acid decomposition or an immune globulin (IG) dosimeter.	Protective clothing for chemical, biological, radiological and nuclear defense that warns of exposure to these hazards by using at least one sensor and changing its electrical properties. [29]
b) Schematic sketch		
c) Known/possible field of application	Sensor application to swimwear, leisurewear and workwear for outdoor activities for detection of UV radiation.	Personal protective equipment (PPE).
d) Possible sensor variants	Determination of the intensity of UV radiation by measurement using iodine method, acetone decay, oxalic acid decomposition or IG dosimeter.	Sensor element can be formed from: temperature sensors, pressure sensors, humidity sensors, pH sensors, radiation sensor.
e) Opportunities and challenges	+ Good resistance of the textile carrier to the hazard potential, sensitization technology and reactions causing color change − Doubts as to whether the concentration and intensity of the hazard potential is sufficient to initiate the chemical reaction on the textile	+ Indication of the end of the recommended wearing period + No time limit for the wearing period
I MATERIAL PROPERTIES		
II ENERGY SUPPLY	Light	Electric current
III RESOLUTION		
IV SENSITIVITY		
V MEASUREMENT RANGE		
VI TRL	<6	6–8

	CLOTHTECH	
	PHOSPOHR TEMPERATURE SENSOR	**PHYSIOLOGICAL SENSOR 1**
SENSOR TYPE	Thermal	Chemical, mechanical
MEASURAND	Temperature	Electric current
CONSTRUCTION PRINCIPLE	Weave	Conductive yarn
GEOMETRY	Planar	Planar
MATERIAL	Aramid fibers	Silver/gold-coated nylon
Procedural principle	Multilayer garment whose outer sheath is equipped with a temperature sensor. This ensures rapid expansion of the textile under the effect of heat to ensure a heat-insulating intermediate layer to protect the body. As material, aramid fibers like MPD-I, PPD-T, PBI are used. [30]	Recording of physiological states via sensors, transmission via electrically conductive conductor paths in clothing and processing in measuring equipment. [31]
Schematic sketch	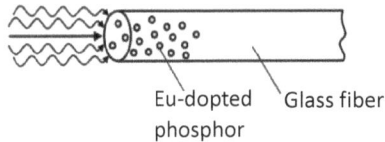 Eu-dopted phosphor Glass fiber	
Known/possible field of application	Protective clothing (coat, jacket, trousers) for firefighters or industrial applications under high heat exposure.	Sportswear and medical clothing for monitoring bodily functions. Multimedia clothing for adapting media enjoyment to physiological conditions.
Possible sensor variants	A 25% increase in time until second-degree burn occurs on the skin compared to conventional protective clothing. Sportswear/medical clothing. [32], [33], [34], [35], [36]	Textile electrode in spacer warp-knit. [37], [38], [39] Multifunctional apparel system. [40]
Opportunities and challenges		+ Advantageous contact behavior due to pressure-elastic behavior when using monofilaments + Acceptance by the wearer due to attractive appearance + Comfortable to wear due to the flexibility of the garment
MATERIAL PROPERTIES	Riskiness from 1.5 to 4	Electrical resistance: <5 Ω/cm; diameter of monofilaments: >100 µm
ENERGY SUPPLY	Heat	Electric current
RESOLUTION	0–3 s	
SENSITIVITY		
MEASUREMENT RANGE		
TRL	9	9

CLOTHTECH

	PHYSIOLOGICAL SENSOR 2	PHYSIOLOGICAL SENSOR 3
1 | SENSOR TYPE	Mechanical	Thermal
2 | MEASURAND	Electromagnetic light spectrum	Electromagnetic light spectrum
3 | CONSTRUCTION PRINCIPLE	Fiber, fiber braid, diffraction grating	Weft knit, warp knit, weave
4 | GEOMETRY	Optical fiber with outer diameter of 125 µm, grating diameter of 6–9 µm, sensor diameter of 150–250 µm	Sheath diameter: 0.125 mm; core diameter: 0.09 m
5 | MATERIAL	Polymers, glass, electrically conductive metals	
a) Procedural principle	A patient-monitoring system comprising a plurality of diffraction gratings arranged along an optical fiber. Each optical fiber and grating is configured to change either the effective refractive index or the grating periodicity of the corresponding grating at its location along the fiber in response to at least one desired external stimulus. [41]	Integration of a fiber-optic temperature sensing element into a fabric. The temperature sensing element is an optical fiber containing one or more fiber-Bragg-grating sensors. Light is introduced into the optical single-mode fiber and directed to a grating interface adjacent to the wearer. A reflux signal is received by a reflection mode or a transmission mode, the reflux signal having a wavelength shift that is indicative of temperature v the Bragg resonance effect. [42]
b) Schematic sketch	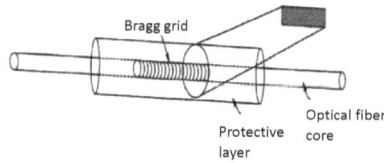	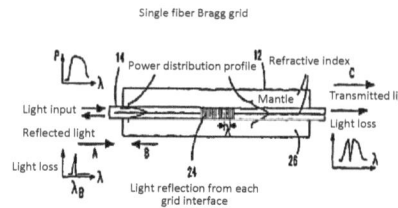
c) Known/possible field of application	Nursing of newborns.	
d) Possible sensor variants	Reduced number of required connection options.	Processing of the thread in a weft knit, warp knit or weave.
e) Opportunities and challenges	+ Hygiene + Skin sensitivity + Wearing acceptance	
I MATERIAL PROPERTIES		
II ENERGY SUPPLY	Electric current	Electric current
III RESOLUTION		
IV SENSITIVITY		
V MEASUREMENT RANGE		
VI TRL	9	6–8

CLOTHTECH

	PHYSIOLOGICAL SENSOR 4	PHYSIOLOGICAL SENSOR 5
SENSOR TYPE	Mechanical, chemical, thermal	Mechanical
MEASURAND	Electric current	Electric current
CONSTRUCTION PRINCIPLE	Elastic weave or fleece containing electrically conductive fibers	Weft knit containing electrically conductive threads
GEOMETRY	Linear, planar	Punctiform, linear or planar
MATERIAL	Elastomers filled with conductive particles or electrically conductive metals	
Procedural principle	Garment with belts running transversely to the longitudinal axis of the wearer, which can be stretched in the longitudinal direction and in which strain gauges are incorporated, which allow physiological functions to be determined by changing the electrical conductivity. [43]	Sensor consisting of strain gauges, piezoelectric elements, length gauges or pressure sensors, all of which change their electrical properties under mechanical deformation. [44]
Schematic sketch		
Known/possible field of application	Clothing for monitoring heart activity and recording skin resistance, perspiration and body temperature.	Garment for determining a posture or movement of the body.
Possible sensor variants	The carrier material of the electrically conductive threads is knitted fabric made of cotton with elastane content or viscose, or synthetic or microfiber. Conductive particles in the strain sensor can be carbon particles or hydrogels.	Sensor element can be formed from strain gauges.
Opportunities and challenges	+ The garment should be resistant to perspiration and washing + Increase of sensor sensitivity through path-shaped guidance of the strain sensor, since the transverse elongation is low compared to a longitudinal elongation − An insulating layer should prevent moisture from influencing the measurement signal of the extensometers − The elastomer should be more extensible than the substrate on which the sensor is placed so that the extensibility of the sensor does not limit that of the garment	+ Piezoelectric elements + Magnetic, capacitive or optical length gauges. + High wearing comfort due to unobtrusive integration of the sensor elements into the garment
MATERIAL PROPERTIES	Specific sensor resistance 5–30,000 Ωcm	
ENERGY SUPPLY	Electric current	Electric current
RESOLUTION		
SENSITIVITY		
MEASUREMENT RANGE		
TRL	9	9

	CLOTHTECH	
	TENSILE STRAND	**PHYSIOLOGICAL STRAIN SENSOR**
1 | SENSOR TYPE	Mechanical	Mechanical
2 | MEASURAND	Electric current	Electric current
3 | CONSTRUCTION PRINCIPLE	Weft knit, warp knit, weave	Elastic fabric containing electrically conductive filaments
4 | GEOMETRY	Linear, planar	Linear, planar
5 | MATERIAL	Elastic, electrically conductive core thread; bimetallic sheathing	
a) Procedural principle	Thread for determining the tensile stress which consists of an elastic core thread and has at least one sheath, this sheath changing its electrical property, in particular its electrical resistance and/or its capacitance, when the length of the core thread changes. As a result of the tensile stress, pressure is exerted on the body of the wearer. [45]	Electrically conductive thread for determining the state of respiration and movement, which changes its electrical properties under tensile or compressive load, above all its electrical resistance and inductance. [46]
b) Schematic sketch	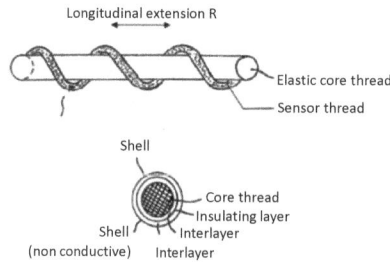	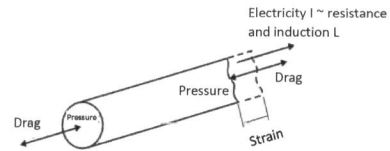
c) Known/possible field of application	Bandage or compression stocking. Sheathing can release substances to the skin of the wearer.	Clothing for monitoring respiratory and physical activity of newborns, children, adults and even non human mammals.
d) Possible sensor variants	Processing of the thread—preferably as weft thread—in a weft knit, warp knit, or weave.	Clothing for monitoring respiratory and physical activity of newborns, children, adults and even non human mammals.
e) Opportunities and challenges	+ Single- or multilayer sheathing + Thread with one or more wrapping threads	+ Moisture resistant, i.e., washable + High measurement accuracy + No slipping of the sensors, due to precise positioning in the garment + Increased wearing comfort due to the not tootight fit of the garment
I MATERIAL PROPERTIES		
II ENERGY SUPPLY	Electric current	Electric current
III RESOLUTION		
IV SENSITIVITY		
V MEASUREMENT RANGE		
VI TRL	9	9

	CLOTHTECH	
	PRESSURE SENSOR	**KNITTED BREATHING SENSOR**
SENSOR TYPE	Mechanical	Mechanical
MEASURAND	Electric current, pressure	Electric current
CONSTRUCTION PRINCIPLE	Elastic weft-knit or warp knit containing electrically conductive threads	
GEOMETRY	Dimensions of the pressure sensor: 10 mm x 10 mm x 1 mm	
MATERIAL	Electrically conductive metals	Electrically conductive polyester with stainless steel content
Procedural principle	Pressure sensor with strip-like or filament-like elements which each have a layered structure and are electrically conductive. When pressure is applied, the layers touch each other and a closed circuit is formed which indicates the pressure. [47]	Measurement of respiratory movement by changing the electrical resistance when weft knits made of conductive polyester fiber yarns with stainless steel content are stretched. [48]
Schematic sketch		
Known/possible field of application	Clothing for monitoring heart activity and recording skin resistance, perspiration and body temperature.	
Possible sensor variants	Pressure-sensitive stocking. [49]	Right/left weft knit with conductive stripes. Right/left lining weft knit in which the conductive yarn no longer forms any stitches, but is merely integrated with handles in the non-conductive basic knit Right/right weft knits where the electrical resistance is less dependent on elongation
Opportunities and challenges		
MATERIAL PROPERTIES	Environmentally stable at 0–50 °C, 30–90% relative humidity	
ENERGY SUPPLY	Electric current	Electric current
RESOLUTION		
SENSITIVITY		
MEASUREMENT RANGE	Surface pressure: 0–10 kg/cm^2	
TRL	6–8	6–8

Schematic sketch labels: Electricity I ~ resistance R and induction L; Drag; Pressure; Strain; Drag; Pressure

	THREE-DIMENSIONAL SPACER WARP KNIT	SHAPE-MEMORY SENSOR
1 | SENSOR TYPE	Mechanical	Thermal
2 | MEASURAND	Electric current	Electric current, temperature
3 | CONSTRUCTION PRINCIPLE	Warp knit	Wire, integrated in support fabric
4 | GEOMETRY	Planar	Planar
5 | MATERIAL		Metallic alloys, polymers
a) Procedural principle	Three-dimensional spacer warp-knit with integrated ultrasonic sensors for monitoring body movement. [50]	By heating to a certain temperature via an electric current, the fabric takes on a desired shape with integrated conductive wires. When the electric current is deactivated, the material returns to its original shape. [51]
b) Schematic sketch	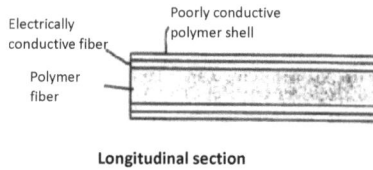 Electrically conductive fiber / Poorly conductive polymer shell / Polymer fiber / **Longitudinal section**	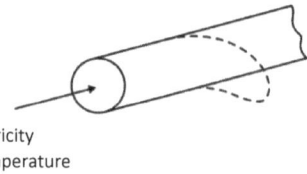 Electricity ~ temperature
c) Known/possible field of application		Protective suit for pilots of fighter planes who are exposed to high forces on the body due to large accelerations.
d) Possible sensor variants	Flexible elastane material guarantees flexibility and wearing comfort.	Disposable shape-memory effect: by only one phase transition in the metallic alloy, the material can only reach its original state. Two-way shape-memory effect: two different original material states can be achieved by varying the temperature into a high and a low temperature.
e) Opportunities and challenges		+ Fast reaction time + Functional maintenance even with minor damage + The total weight and installation space of the device are less than those corresponding to the state of the art − Permanent irreversible plastic deformation of up to 0.1%
I MATERIAL PROPERTIES		
II ENERGY SUPPLY	Electric current	Electric current
III RESOLUTION		0.2–1 s
IV SENSITIVITY		
V MEASUREMENT RANGE		<100 °C
VI TRL	6–8	9

	CLOTHTECH	
	SAFETY CLOTHING	**INTELLIGENT SKIN ARCHITECTURE**
SENSOR TYPE	Mechanical	Mechanical
MEASURAND	Electric current	
CONSTRUCTION PRINCIPLE	Weft knit containing electrically conductive threads	Weave
GEOMETRY	Linear, planar	Planar
MATERIAL	Electrically conductive metals	
Procedural principle	Garment for locating stab wounds or gunshot wounds to the human body by using sensor units arranged in electrical conductor tracks, the operating principle of which is the piezoelectric effect. [52]	Optical fibers woven into a carrier material which serve as sensors for optical information transmission. [53]
Schematic sketch		
Known/possible field of application	Protective vests for the police and military.	Acquisition of data; image processing; communication.
Possible sensor variants	Variation of the conductor arrangement, preferably in wave or curve form, as these ensure an elastic arrangement. The construction of many smaller circuits enables a more precise location of the interruption of the conductor path and thus of the injury to the wearer. Polymer tracks can be printed, embroidered or woven directly onto the fabric.	Supporting weaving of the optical fibers into channels. Arrangement of the optical fibers in a grid-like mat consisting of fibers of any carrier material. Woven structure comprising a first group of warp-direction yarns and a second group of weft-direction yarns with optical fibers arranged between selected pairs of the first group. Optoelectronic packaging structure with two sections, in each of which the abovementioned woven structure is placed.
Opportunities and challenges	+ Detection of impacts, pressure waves and detonations using piezoelectric elements	+ Low construction volume; low weight + High tensile strength, high elasticity, high resistance to weathering, high resistance to chemicals, high tear strength, high dimensional stability, high wear resistance − Sensitivity to deflection, leading to a deterioration in transmittance
MATERIAL PROPERTIES		
ENERGY SUPPLY	Electric current	None
RESOLUTION		
SENSITIVITY		
MEASUREMENT RANGE		
TRL	9	6–8

	CAPACITIVE BREATHING SENSOR	CARBON NANOTUBE (CNT) STRAIN SENSOR
1 | SENSOR TYPE	Mechanical	Mechanical
2 | MEASURAND	Electromagnetic field	Electric current
3 | CONSTRUCTION PRINCIPLE	Conductive ink between textile layers	Yarn
4 | GEOMETRY	Planar	Linear
5 | MATERIAL	Combination of stretchable and non-stretchable textiles with conductive ink	Carbon nanotube yarn
a) Procedural principle	Respiration measurement via capacitive proximity sensor. Respiratory frequency is measured by the displacement of two textile layers, which are connected by a conductive layer, caused by respiratory movement. [54]	Electrical resistance of twisted CNT yarns changes with change in load or temperature. [55]
b) Schematic sketch		
c) Known/possible field of application		Monitoring of motion and temperature.
d) Possible sensor variants		An advanced strain sensor for human motion detection was introduced by Yamada. It uses a new material, namely thin films of aligned single-walled carbon nanotubes. Unlike traditional rigid materials such as silicon, nanotube films fracture into gaps and islands, and bundles bridge the gaps. This allows the films to function as strain sensors capable of measuring strains of up to 280% with high durability.
e) Opportunities and challenges		
I MATERIAL PROPERTIES		
II ENERGY SUPPLY	Electric current	Electric current
III RESOLUTION		
IV SENSITIVITY		Strain 1.4–1.8 mV/V/1000 m; temperature: 91 mA/°C
V MEASUREMENT RANGE		
VI TRL	<6	<6

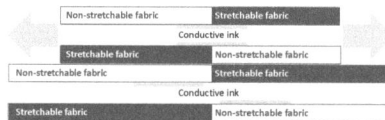

Non-stretchable fabric — Stretchable fabric
Conductive ink
Stretchable fabric — Non-stretchable fabric
Non-stretchable fabric — Stretchable fabric
Conductive ink
Stretchable fabric — Non-stretchable fabric

CLOTHTECH

	BIOPOTENTIAL SENSORS	PRESSURE FORCE MAPPING SENSOR
SENSOR TYPE	Mechanical	Mechanical
MEASURAND	Electric current	Electric current
CONSTRUCTION PRINCIPLE	Weaves, weft knits, embroidered electrodes	Weave, weft knit
GEOMETRY	Planar	Planar
MATERIAL	Silver yarns for electrodes	Carbon black, metal, and metal oxide particles
Procedural principle	Electrocardiography (ECG) and electromyography (EMG) are the electrical potentials periodically changed by cardiovascular and muscle activities. [23]	The technology behind force mapping is typically a grid of individual force sensor elements. The core principle of electrical resistance-based pressure mapping is the special property of electrically conducting polymer composites (ECPC), i.e., that their deformation, which could be caused by either tension or pressure, will cause their electrical impedance in the vicinity of the deformation to change. [56]
Schematic sketch		
Known/possible field of application		
Possible sensor variants	Nervous stimuli and muscle contraction can be easily detected by measuring the ionic current flow in the body. This measurement is accomplished by attaching biopotential electrodes to the skin surface. ECG/EMG-monitoring systems. The electrodes are either made of gel or stuck to the skin using conductive adhesives in order to develop better contact with the skin. To improve contact between the electrodes and the skin, skin preparation is required, such as shaving, abrading and cleaning the skin surface. A wearable electrode is created by weaving, knitting or stitching silver yarns on the inner surface of the clothing. Due to their irregular surface structures, this creates high impedance, and therefore high-frequency noise.	Force sensors can be implemented based on various principles such as piezoresistive, piezoelectric, piezomagnetic, capacitive, magnetic and optical. The basic physical structure of capacitive-based pressure-mapping sensors is two parallel conductive plates separated with a flexible, non-conductive layer as the dielectric spacer.
Opportunities and challenges	– Gelatinous substances dry out over a long period of time and cause the electrode to come off the skin. Adhesives can irritate the skin, leading to a loss of signal quality.	+ The sensing elements can be isolated from the skin by either additional regular textile layers or direct isolation coatings to avoid any complications from electrode–skin contact. + Easily scalable in terms of sensing channels. This is mainly because of the simplicity in the measuring structure – Higher data processing/transmission requirements, the need for special conductive and/or dielectric materials, relatively complex sensor structures
MATERIAL PROPERTIES		
ENERGY SUPPLY	Electric current	Electric current
RESOLUTION		40 Hz
SENSITIVITY		
MEASUREMENT RANGE		
TRL	9	6–8

	OPTOELECTRONIC SENSOR	MOISTURE- AND CHEMICAL-SENSITIVE SENSOR THREAD
1 | SENSOR TYPE	Chemical	Mechanical, chemical
2 | MEASURAND	Electromagnetic light spectrum	Visual assessment
3 | CONSTRUCTION PRINCIPLE	Fiber	n/a
4 | GEOMETRY	Linear	Linear
5 | MATERIAL	Plexiglas	n/a
a) Procedural principle	Detection of adhering liquid components in or on liquid-storing substances by detecting the change in the transmission of light in a light guide with the liquid component to be taken. [3]	Permanent identification of harmful environmental influences through the use of threads which change their shape, color or volume while absorbing liquid. The core yarn must be UV-resistant and clearly distinguished in color from the load-bearing tape. For the sheath fibers of the yarn, a material must be selected which is changed in shape, color or structure by UV radiation. [4]
b) Schematic sketch	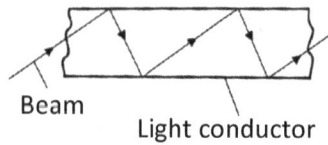 Beam Light conductor	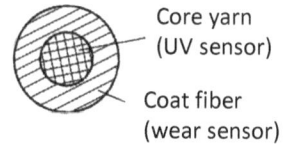 Core yarn (UV sensor) Coat fiber (wear sensor)
c) Known/possible field of application	Detection of liquid content of soils, textiles or granulates. Monitoring tasks, for example in landfills.	A friction-spun sensor thread represents a combination of an abrasion sensor and a UV sensor.
d) Possible sensor variants	Cost-effective.	Decrease in abrasion resistance with increasing exposure to UV radiation.
e) Opportunities and challenges		
I MATERIAL PROPERTIES		
II ENERGY SUPPLY	Light	None
III RESOLUTION		
IV MATERIAL		
V MATERIAL PROPERTIES		
VI TRL	9	9

CLOTHTECH

TEXTILE-BASED GONIOMETER

SENSOR TYPE	Mechanical
MEASURAND	Electric current
CONSTRUCTION PRINCIPLE	Weft knit
GEOMETRY	Planar
MATERIAL	Combination of electroconductive (Belltron® by Kanebo Ltd.) and elastic (Lycra®) yarns

Procedural principle

Knitted piezoresistive fabrics modify their electrical resistance when they are elongated or flexed. The main requirement for the application of the single-layer sensors is that the human movements must produce a strain field which can be detected by resistance variation. For this reason, single-layer sensors must be integrated into adherent garments close to the human joint under investigation. [25]

Schematic sketch

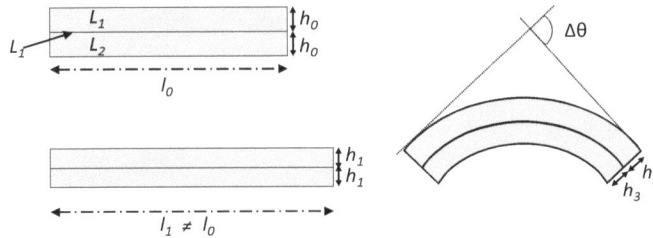

Known/possible field of application

Hand motion sensing: a kinesthetic sensing glove was developed for the ambulatory evaluation of the residual hand function and its recovery in post-stroke patients; scapular movement detection.

Possible sensor variants

Single-layer sensor or double-layer sensors.

Opportunities and challenges

– Needs to closely adhere to joint moving

MATERIAL PROPERTIES	
ENERGY SUPPLY	Electric current
RESOLUTION	
SENSITIVITY	Single layer: 6405 Ω for $\Delta\theta = 37°$
MEASUREMENT RANGE	double layer: 5100 Ω for $\Delta\theta = 37°$
TRL	6–8

	WATER DETECTOR	FIBER-OPTIC SENSOR
1 | SENSOR TYPE	Chemical	Chemical
2 | MEASURAND	Visual assessment	Electromagnetic light spectrum
3 | CONSTRUCTION PRINCIPLE	Textile tape, thread, thread bundle, textile fiber composite, fleece, paper, film, wire, warp knit	Fiber
4 | GEOMETRY	Punctiform, linear, planar, voluminous	Linear
5 | MATERIAL	Cellulose, polyolefin, nylon, Nomex, Teflon, plastic, polyester, ceramic, metal, wool	Cotton for protective vision, fluoride glass for light-guide sheath and core
a) Procedural principle	Textile probe with sufficiently large stored active substance depot, which on contact with the substance to be investigated causes a visual chemical change in the detector depending on the composition and movement of the analyte. The change occurs in the form of a substance solution, substance deposition or formation of a new substance at the detector itself. [5]	Fiber-optic sensor for detecting gaseous or liquid media, surrounded by an optical fiber sheath consisting of a fluoride glass of low chemical resistance to be detected on contact with the analyte. Decomposition of the sheath takes place within a characteristic chemically induced reaction time until the sensor responds as a function of the original thickness of the sheath, the temperature and the concentration of the attacking medium while maintaining the total reflection condition (lower refractive index of the sheath with respect to the optical fiber core). A hygroscopic protective textile layer around the light-guide sheath increase the corrosive effect of the attacking medium on the light-guide sheath. [7]
b) Schematic sketch		Fiber-optic core Fiber-optic sheath Gas- and liquid-permeable protective cover
c) Known/possible field of application	Analysis of gas and water, and also soil and sediment, samples.	Detection of gaseous and liquid media. Monitoring of electrical cables and lines, as well as endangered installations, pipelines, equipment and buildings for the ingress of water, water vapor, acid alkalis or other gases and liquids.
d) Possible sensor variants	The resistance of the optically visually-recognizable color pattern of the detector to water with a different composition to that of the measuring point and the atmosphere, which is exposed to short-term effects, prevents falsification of the measurement.	High mechanical strength.
e) Opportunities and challenges		+ High response sensitivity, even to individual media only + Targeted analysis of individual specific substances with desired concentration content + Low manufacturing and general cost
I MATERIAL PROPERTIES		Light-guide sheath with lower refractive index than the conductor core, light guide sheath made of fluoride glass with lower hydrolytic resistance
II ENERGY SUPPLY	None	None
III RESOLUTION	>1 h	
IV MATERIAL		
V MATERIAL PROPERTIES		
VI TRL	9	6–8

GEOTECH

	CARBON-FILLED CELLULOSE PHASE	PH SENSOR
SENSOR TYPE	Mechanical	Chemical
MEASURAND	Electric current	Visual assessment
CONSTRUCTION PRINCIPLE	Fiber, filament, film	
GEOMETRY	Linear, planar	Linear
MATERIAL	Polymer	
Procedural principle	Carbon-filled cellulose fiber. Detection of liquids or vapors via electrically conductive filaments from dry-wet spun cellulose dotted with charge carriers (graphite, carbon black, pigments with semiconducting layers, metallic fibers or carbon fibers) whose conductivity changes under tension/pressure or with increasing moisture content. [8]	Measurement of substance concentrations, which are not directly accessible spectroscopically, with a sensitive chemoreceptor. This receptor is a sensor, at the end of which a specific indicator (e.g., phenol red in polyacrylamide) is immobilized by which a change in pH is measured either in reflection or as fluorescence. [9]
Schematic sketch		
Known/possible field of application	Detection of liquids or vapors.	
Possible sensor variants	Mechanically stable even at high temperatures. Sometimes even fire-retardant.	Very accurate pH measurement only achievable for very small ranges (approximately three pH units).
Opportunities and challenges	− Increasing carbon-black content reduces substance strength, ductility and toughness − Doping with carbon black influences the material viscosity to such an extent that stable thread formation is not possible at normal spinning speeds − If the doping with soot is too high, the electrical resistance increases disproportionately	
MATERIAL PROPERTIES		
ENERGY SUPPLY	Electric current	None
RESOLUTION		
MATERIAL		
MATERIAL PROPERTIES		0.005 pH units
TRL	6–8	6–8

	FIBER-OPTIC PH SENSOR	INTEGRATED OPTICAL RESONATOR
**1	SENSOR TYPE**	Chemical	Thermal
**2	MEASURAND**	Electromagnetic light spectrum	Electromagnetic light spectrum
**3	CONSTRUCTION PRINCIPLE**	Fiber	
**4	GEOMETRY**	Linear	Linear
**5	MATERIAL**	Polymer, glass	$LiNbO_3$
a) Procedural principle	Utilization of the light absorption dependent on the pH value of the surrounding medium in a fiber-optic probe consisting of a segment of a multimode optical fiber whose end forms the sensor head. In this area, both the coating and the cladding of the fiber are removed, so that a sensitive layer of a copolymer with immobilized dye is polymerized onto the core. Electromagnetic radiation is guided in such a way that the light rays pass through the interface between the fiber core and the sensitive layer and are returned to the core by total reflection at the interface between the sensitive layer and the aqueous analyte. Wavelength-selective absorption occurs. [10]	The temperature changed by means of optical resonators integrated in LiNbO3 with a periodic characteristic curve. In order to be able to record the number of orders passed as a function of the direction of the phase (or temperature) change, it requires two signals phase-shifted by 90°. It is advantageous to use the output signals to arrive at an evaluation, which counts in each case with the zero crossing, and thus an independence from slow fluctuations of the light intensity is obtained. The phase modulation required for differentiation is achieved by frequency modulation of the laser light or by electro-optical modulation of the optical path length of the resonator. [1]
b) Schematic sketch		

Cladding Coating pH-sensitive layer Mirror layer
Protective layer
6–20 mm
Unmirrored face Fiber core Shaft Epoxy resin

Light
c-Axis
$TiLiNbO_3$
$LiNbO_3$

c) Known/possible field of application	Chemical-analytical measurements.	Temperature monitoring of textile structures.
d) Possible sensor variants	Low influence of the internal thickness on the sensor characteristic curve.	The sensitivity of the temperature sensor can be determined in wide ranges by the length of the component and the wavelength of the light.
e) Opportunities and challenges	+ High long-term stability + High sensitivity + Damping arm	+ Simple measuring system with high accuracy when supplying the resonator sensor element vi a polarization-maintaining monomode fiber + Measurement of smallest temperature changes possible due to the strong temperature dependence of the refractive index − Measurement of absolute temperatures not possible
I MATERIAL PROPERTIES	Six months service life	
II ENERGY SUPPLY	Light	None
III RESOLUTION		
IV MATERIAL		Sensitivity of 35 impulses/K, resolution of 29 impulses/K
V MATERIAL PROPERTIES	At 680 nm, 0.06 absorbance units per pH unit over the measuring range of four pH units	
VI TRL	9	6–8

	GEOTECH	
	TEMPERATURE SENSOR	**PHOSPHOR TEMPERATURE SENSOR**
SENSOR TYPE	Thermal	Thermal
MEASURAND	Electric current, electromagnetic light spectrum, transmitted light, temperature	Electromagnetic light spectrum, temperature
CONSTRUCTION PRINCIPLE	Thread	Thread
GEOMETRY	Linear	Linear, phosphorus diameter a few 100s of µm
MATERIAL	Metals, electrically conductive polymers, glass fibers	Doped phosphorus (Gd_2O_2S and La_2O_2S)
Procedural principle	Design of thread-shaped sensors for the investigation of thermal loads based on low-melting metal wires, which change their electrical properties under thermal load. [4] Temperature determination by measuring the change of the refraction coefficient of the light-guide sheath under temperature change, which leads to a corresponding transmission difference. [9]	Temperature determination with evaluation of the temperature-dependent luminescence of a doped phosphor at the end of an optical glass fiber to generate a luminescence, the phosphor is excited by UV light via a (multimode) fiber and the fiber guided over the same fiber is spectrally decomposed and detected. The intensity ratio of two lines determines the temperature. [1]
Schematic sketch	Heat Low-melting metal	Eu-dopted phosphor Glass fiber
Known/possible field of application	Temperature monitoring of textile structures.	Temperature monitoring of textile structures.
Possible sensor variants	Use of threads of electrically conductive polymers or electrically conductive coated polymers. Temperature sensors based on the principle of absorption edge displacement using filter glasses instead of semiconductor elements.	As an alternative to the intensity ratio, the temperature-dependent phase shift between luminescent light and excitation light can be determined with periodic excitation. The measurement range of this variant is between -30 and 150 °C with an accuracy of 0.04 °C. Using a small, inexpensive and luminescent GaxAl1-x- As crystal as a sensor, a temperature range between 0 and 200 °C can be measured with an accuracy of 1 °C (resolution 0.1 °C).
Opportunities and challenges	+ High reproducibility + Short response time + High accuracy + Low tendency for thread or surface production due to unfavorable properties of the metals	+ Cost-effecitve + Small installation space
MATERIAL PROPERTIES		
ENERGY SUPPLY	None	None
RESOLUTION		
MATERIAL		0.1 °C
MATERIAL PROPERTIES	50–250 °C	-50–+250 °C
TRL	6–8	6–8

	GEOTECH	
	FIBER-OPTIC DISPLACEMENT TRANSDUCER	**ACTIVE FIBER-OPTIC SENSOR**
1 | SENSOR TYPE	Mechanical	Chemical
2 | MEASURAND	Path, route	Visual assessment
3 | CONSTRUCTION PRINCIPLE	Fiber-optic conductor	Fiber
4 | GEOMETRY	Linear	Linear
5 | MATERIAL		
a) Procedural principle	Measurement of paths on the basis of various principles. In particular, fiber optic measurements of a large number of physical quantities that can be converted into paths by test specimens. [1]	Measurement of the distance between sensor and fluid environment, the concentration of chemicals in the fluid environment, the pH value of aqueous solutions, and the partial pressures of a gas by evaluating the light transmitted via the fiber-optic laser if this changes characteristically as a reaction between sensor reagent and surrounding environment. [12]
b) Schematic sketch		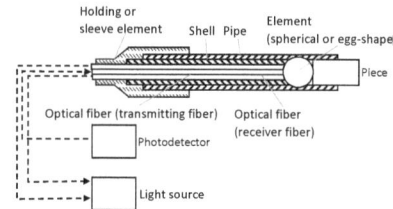
c) Known/possible field of application	Measurement technology, from displacement measurement, angle, pressure or acceleration can also be measured, depending on the arrangement.	Control of chemical processes in nuclear and industrial areas, underground nuclear waste in the environment, in medical and biological analysis, as well as in the agri-food industry; medical applications; biochemical applications; use in the food industry.
d) Possible sensor variants		Fiber-optically active sensor. [13]
e) Opportunities and challenges		+ Long service life + Simple sterilization + High stability – Limited pH measuring range – Limited reproducibility of the reaction between optical fibers and the immobilized reagent
I MATERIAL PROPERTIES		Bulky sensor material
II ENERGY SUPPLY		Light
III RESOLUTION		
IV MATERIAL		
V MATERIAL PROPERTIES	10^{-10}–1 m	
VI TRL	9	9

	GEOTECH	
	SOUND SENSOR (HYDROPHONE)	**RAPID-SHRINK FIBER**
SENSOR TYPE	Mechanical	Chemical
MEASURAND	Electric current	Visual assessment
CONSTRUCTION PRINCIPLE	Fiber-optic conductor	Warp knit, weave
GEOMETRY	Linear	Linear, planar
MATERIAL	Quartz glass	Elastomer (polyutherane, rubber)
Procedural principle	Fiber optic hydrophone (Mach–Zehnder interferometer) for highly sensitive detection of pressure differences between measuring and reference fibers. By modulating the refractive index of the measuring fiber, the sound pressure changes the phase length of the passing light and thus the interference signal, which is detected by two photodiodes and fed to the amplifier via a high-pass filter. The signal behind the low pass is used to stabilize the operating point of the interferometer against slow fluctuations, e.g., due to temperature changes. [1]	A polymer fiber which shrinks rapidly at ordinary temperature and in contact with water, but retains its shape (impact strength), has high absorbency, and has performance characteristics such as rubber elasticity. [14]
Schematic sketch		
Known/possible field of application	Metrology.	Disposable diapers; fastening tapes; cloths as covers for dampening units in offset printers; cords or cylinders for plant cultivation; cords and nets for the food industry; bank reinforcements.
Possible sensor variants	Due to the flexibility of the quartz glass fibers, sensors with directional characteristics can be manufactured.	A water-absorbing, shrinkable yarn produced by blending or by blending spinning the rapidly shrinking fiber and a fiber that shrinks slower than said fiber upon absorption of water. A water-absorbing shrinkable material which consists of a water-absorbing shrinkable fibrous web and a water-absorbing shrinkable yarn that absorbs water at a higher rate and to a greater extent than the fibrous web, with the water-absorbing shrinkable yarn containing the rapidly shrinking fiber.
Opportunities and challenges		
MATERIAL PROPERTIES		At 20 °C, maximum percentage shrinkage >30%
ENERGY SUPPLY	Laser light	
RESOLUTION		0–10 s
SENSITIVITY		
MEASUREMENT RANGE		At 20 °C, shrinkage stress = 0.351–1.755 kg/m^2 (30–150 mg/den)
TRL	9	9

	PYROMETERS	POLYIMIDE WAVE CONDUCTOR
1 | SENSOR TYPE	Thermal	Chemical
2 | MEASURAND	Electromagnetic light spectrum, temperature	Moisture content
3 | CONSTRUCTION PRINCIPLE	Fiber-optic conductor	Fiber
4 | GEOMETRY	Linear	Planar
5 | MATERIAL	Sapphire glass, quartz glass	Polyamide-imide, respectively perfluorinated polyimide, substrate
a) Procedural principle	Fiber-optic measurement method that determines the temperature by analyzing the cavity radiation of a black body. The radiation spectrum of the black body shifts according to Planck's law of radiation depending on temperature. [1]	Optical sensor for the quantitative determination of liquids in the vapor phase, comprising a cover layer, one or two layers of a polyamide imide or a perfluorinated polyimide, and a substrate. [57]
b) Schematic sketch	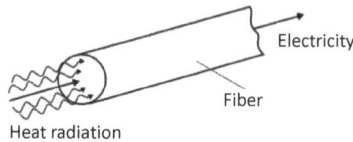	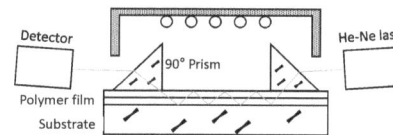
c) Known/possible field of application	Non-contact temperature measurement.	Determination of polar and non-polar liquids in the vapor phase as well as NH3, NH4OH, NO2 and N2O5 by exploiting the sensitivity of polyimides to moisture due to interactions with liquid componen near the surface.
d) Possible sensor variants	Very small heat capacity allows measurement of rapid temperature changes.	Formation of the polyimide waveguide as strip waveguide, interferometer structure, directional coupler structure.
e) Opportunities and challenges	+ Measurement of very high temperatures possible	+ Functionality for polar and non-polar liquids + Functionality in vacuum + Independent measurement method against fluctuations in absolute values, since a comparative measurement of the phase differences of two polarizations takes place + Digital evaluation possible − Differentiation between water and other liquids
I MATERIAL PROPERTIES		
II ENERGY SUPPLY	Electric current	
III RESOLUTION		
IV SENSITIVITY	Measurement accuracy of 0.05%	
V MEASUREMENT RANGE	Up to about 2000 °C	
VI TRL	9	9

	GEOTECH	
	GYROSCOPE (ROTATION SENSOR)	**INTENSITY-BASED FIBER-OPTICAL SENSORS**
SENSOR TYPE	Mechanical	Chemical
MEASURAND	Electric current	Electromagnetic light spectrum
CONSTRUCTION PRINCIPLE	Fiber optic	Fiber
GEOMETRY	Linear conductor, fiber length between 100 and 1000 m	Linear
MATERIAL		Optically conductive fibers
Procedural principle	Ring interferometer which evaluates the phase difference between the opposing light waves (Sagnac effect), which is dependent on the angular velocity, as a measured variable. Polarized laser light passes between two beam splitters before it is coupled into the two ends of the same fiber coil. In the case of a stationary system, light paths of equal lengths of the circulating modes result in a constructive interference at the output of the second beam splitter, whereas a destructive interference occurs at the output of the first beam splitter. The relativistic Sagnac effect results in a phase difference $\Delta\Phi$ between the light waves rotating in opposite directions, which is proportional to the product of the conversion number m and the enclosed area A. [1]	A measurand-induced change in the optical intensity propagated by an optical fiber can be produced by different mechanisms, such as micro-bending loss, attenuation, and evanescent fields. Requires lighter fibers. They usually use multi-mode large-core fibers. [24]
Schematic sketch	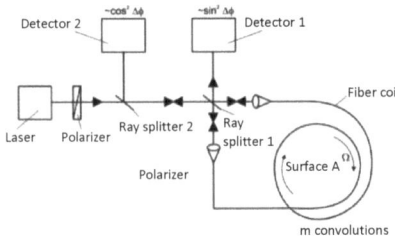	
Known/possible field of application	Earth rotation measurement. Navigation tools. Robot control.	
Possible sensor variants	Integrated optical resonator: sensitivities up to several 100s of °/h.	
Opportunities and challenges	+ Miniaturization of the fiber-optic gyroscope through integrated optics + Use in areas with short-term stability as well as with required long-term stability possible	+ Advantages of this category are easy implementation, low cost, multiplexing and the possibility to implement distributed sensors − Disadvantages include the measurements and variations in the intensity of the light source, which could lead to false readings if a reference system is not used
MATERIAL PROPERTIES		
ENERGY SUPPLY	Laser light	Light
RESOLUTION		
SENSITIVITY	Up to 3–10 °/h	
MEASUREMENT RANGE		
TRL	9	6–8

Schematic sketch labels: Detector 2, Detector 1, ~cos² Δφ, ~sin² Δφ, Fiber coil, Laser, Polarizer, Ray splitter 2, Ray splitter 1, Polarizer, Surface A, m convolutions

	GEOTECH	
	INTRINSIC DISTRIBUTED FIBER-OPTIC SENSORS	**FBG SENSORS**
1 | SENSOR TYPE	Chemical	Chemical
2 | MEASURAND	Electromagnetic light spectrum	Electromagnetic light spectrum
3 | CONSTRUCTION PRINCIPLE	Fiber	Fiber
4 | GEOMETRY	Linear	Linear
5 | MATERIAL	Optically conductive fibers	Optically conductive fibers
a) Procedural principle	Based on Rayleigh scattering. The light is subjected to attenuation due to this scattering, which is determined by random microscopic variations. If a narrow optical pulse is launched in the fiber, it is possible to determine the spatial variations in the fiber scattering coefficient or the attenuation by monitoring the variation of the Rayleigh backscattered signal intensity. The scattering coefficient of a location is influenced by the local fiber status. [24]	FBGs are characterized by periodic changes create by an intense interference pattern of UV energy in the index of refraction in the core of a single-mode optical fiber. The grating reflects a spectral peak based on the grating spacing; therefore, a variation in the length of the fiber due to tension or compression determines a change in the grating spacing, and consequently of the wavelength of light that is reflected. By measuring the center wavelength of the reflected spectral peak, it is possible to obtain a quantitative measurement of the strain. [24]
b) Schematic sketch		
c) Known/possible field of application		
d) Possible sensor variants	Raman scattering is a phenomenon which involves the inelastic scattering of photons. The incident light pulse causes molecular vibrations in the optical fiber. In the case of optical time-domain reflectometry (OTDR), a high input power is requested, as the Raman scattering coefficient is about three orders of magnitude lower than the Rayleigh scattering coefficient. Brillouin scattering is caused by the acoustic vibrations that occur in the optical fiber when an optical pulse is launched. OTDR in different approaches: OTDR based on Rayleigh scattering, OTDR based on Raman scattering, OTDR based on Brillouin scattering. Raman scattering used for the development and implementation of reliable distributed temperature sensors. Rayleigh scattering used to track and to reveal propagation effects.	The response of several FBG sensors can be measured simultaneously by placing several networks in series attached to one lead optical fiber. This is a relevant advantage with respect to traditional strain sensor measurement, which requires an acquisition system for each sensor. By using different wavelengths that are reflected, various FBG sensor signals can be identified, and therefore the space-distributed sensors are identified and distinguished. An optical switch mus then be used to connect several optical fibers to the light source and the spectrometer that measures th reflected wavelengths. The direct embedding of optical fibers with FBG in the epoxy resin of fiber-reinforced polymer (FRP) materials allows exact strain measurement in the material. Therefore, the epoxy resin is an effective protection for the optical fiber. Used for quasi-distributed measurement of strain.
e) Opportunities and challenges		
I MATERIAL PROPERTIES		
II ENERGY SUPPLY	Light	Light
III RESOLUTION		
IV SENSITIVITY		
V MEASUREMENT RANGE		
VI TRL	6–8	6–8

	INDUTECH	
	INTEGRATED OPTICAL FREQUENCY DOUBLER	WEAR SENSOR
SENSOR TYPE	Thermal	Mechanical
MEASURAND	Electromagnetic light spectrum	Visual assessment
CONSTRUCTION PRINCIPLE	Fiber-optic conductor	Thread
GEOMETRY	Linear	Linear
MATERIAL		
Procedural principle	Determination of absolute temperatures by means of optical frequency doubling, in which a special light wavelength is required for a known temperature of the resonator in order to achieve a frequency conversion (phase matching of fundamental and harmonic wave) with high efficiency. [1]	Visual assessment of wear by binding colored threads under the fabric surface of tapes and ropes. If wear occurs, the colored threads become visible on the surface. [4]
Schematic sketch	frequency doubler	Colored threads below the textile surface
Known/possible field of application	Temperature monitoring of textile structures.	Structural health monitoring of ropes.
Possible sensor variants	Particularly high efficiency.	
Opportunities and challenges	– The prerequisite for measurement is a tunable, coherent light source with enough power to operate the resonator	
MATERIAL PROPERTIES		
ENERGY SUPPLY	Electric current	
RESOLUTION		
MATERIAL		
MATERIAL PROPERTIES		
TRL	6–8	9

INDUTECH

	FIBER-OPTIC DISPLACEMENT TRANSDUCER	LUMINOUS SIGNAL FILAMENT
1 | SENSOR TYPE	Mechanical	Chemical
2 | MEASURAND	Path, route	Visual assessment
3 | CONSTRUCTION PRINCIPLE	Fiber-optic conductor	Friction-spun yarn
4 | GEOMETRY	Linear	Linear
5 | MATERIAL		Polypropylene core, polypropylene or polyester jacket
a) Procedural principle	Measurement of paths on the basis of various principles. In particular, fiber optic measurements of a large number of physical quantities that can be converted into paths by test specimens. [1]	Friction-spun yarn or wrap-around yarn with a light-intensive signal thread (with color and light effects) visibly integrated into the core from the outside for the detection of a wear condition. The signal thread is covered by a cover sensitive to environmental influences (abrasion, UV radiation, chemicals), which is why this is visually recognisable after exceeding a limit load that is adjustable via the resistance of the cover. [4]
b) Schematic sketch		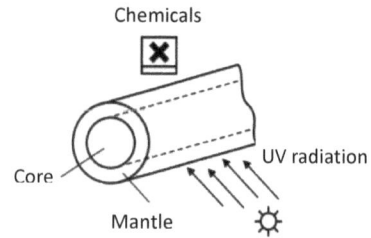
c) Known/possible field of application	Measurement technology, from displacement measurement, angle, pressure or acceleration can also be measured, depending on the arrangement.	Inspection of the wear condition of belts and ropes by means of camera observation.
d) Possible sensor variants		
e) Opportunities and challenges		
I MATERIAL PROPERTIES		Sensitive sheath, fluorescent core
II ENERGY SUPPLY		
III RESOLUTION		
IV SENSITIVITY		
V MEASUREMENT RANGE	$10^{-10}-1$ m	
VI TRL	9	9

INDUTECH

	LAMELLA	HYBRID ROPE
SENSOR TYPE	Mechanical	Mechanical
MEASURAND	Electromagnetic light spectrum	Electromagnetic light spectrum
CONSTRUCTION PRINCIPLE	Composite material; weave, warp knit, weft knit, netting, scrim	Fiber
GEOMETRY	Flat, lamella 400 mm x 200 mm	Linear
MATERIAL	Optical fiber: polymer or glass; carrier textile: glass, carbon, aramid, or basalt scrim	Protective layer: cotton; light-guide sheath and core: fluoride glass
Procedural principle	Embroidered arrangement of high-performance optical fibers with integrated fiber-Bragg-gratings on a lamella for the detection of temperature changes, elongations, compressions and oscillations in supporting structures. Guiding the fiber-optic sensor in the direction of the lines of force for the detection of tensile, compressive and shear forces and also transversely to the lines of force for temperature compensation. Solidification of the textile structure by means of a resin system and construction of the composite material from one or more textile layers. [16]	Investigation of the wear condition on the load-bearing rope by evaluating the ratio of the refractive index between rope core and sheath. The rope is a composition of several modules of different properties, at least one module A having the primary load-bearing function and the secondary driving function and one module B having the primary driving function and the secondary load-bearing function. By inserting conductive elements into the non-conductive modules and sensors, rope elongation can be measured by determining the position of a counterweight. [18]
Schematic sketch	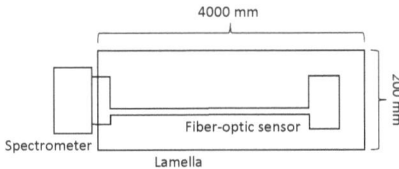	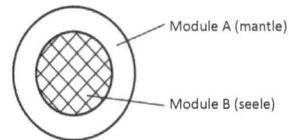
Known/possible field of application	Reinforcement and monitoring of concrete and wooden structures. Critical deflection of structural elements. Critical crack formation. Evidence of functionality, reliability and safety evidence for remaining useful life.	Structural health monitoring of ropes.
Possible sensor variants	Incorporation of fiber Bragg gratings before or after textile processing.	Targeted analysis of individual specific substances with desired concentration content.
Opportunities and challenges	+ Fast and reliable application for building refurbishments + No temperature dependence	+ High response sensitivity, even to individual media only + High mechanical strength + Low manufacturing and general cost
MATERIAL PROPERTIES		Light-guide sheath with a lower refractive index than the conductor core, light-guide sheath made of fluoride glass with lower hydrolytic resistance
ENERGY SUPPLY		
RESOLUTION		
SENSITIVITY		
MEASUREMENT RANGE		
TRL	9	6–8

	STRAIN SENSOR	CONTROL TEAR STRIP
1 | SENSOR TYPE	Mechanical	Mechanical
2 | MEASURAND	Electric current	Visual assessment, electric current
3 | CONSTRUCTION PRINCIPLE	Thread	Thread
4 | GEOMETRY	Linear, diameter 0.5–2.5 mm	Linear
5 | MATERIAL	Kevlar, carbon-black-filled silicone rubber	Polyester; silver-plated polyamide; metallic fine wires; cellulose fiber filled with carbon; glass
a) Procedural principle	Measuring arrangement for determining the strain state in ropes. Based on the location of metal balls incorporated at defined distances by electro-magnetic means, the strain results from the distance and the traversing speed of the balls, since these variables are associated with a change in the specific electrical parameters. [4]	Permanent indication of a one-time load overrun of a belt due to the failure of a control tear thread at a defined elongation value which is significantly below the elongation at break of the belt. [4]
b) Schematic sketch		

Distance d — Metal sphere — Speed v — Induction sensor — Textiles carrier band

Tension — Weight breakage

	STRAIN SENSOR	CONTROL TEAR STRIP
c) Known/possible field of application	Detection of individual wire breaks in steel ropes, e.g., in kevlar elevator ropes. Use for in situ monitoring and determination of load cycles.	Structural health monitoring of ropes.
d) Possible sensor variants	Measurement of strains and strain peaks on the basis of a reproducible dependence on strain and electrical resistance, also while maintaining the strain state by plastic deformation. [4]	Non-conductive control tear thread: Consisting of textile materials such as polyester or polyamide, whose geometric integration into the textile load handling attachment is decisive for the elongation of the overall system at which failure occurs. Detection of a few percent can be realized by means of control yarns of non-typical textile elongations such as carbon fiber, glass fiber or Twaron aramid filament yarn.
e) Opportunities and challenges	+ For protection against overloading, it is not necessary for the sensor thread to fail. Exceeding a defined strain state is sufficient for the output of an alarm signal + By also detecting strain peaks, strain sensors open up a wide range of applications, from crack sensors to sensors for detecting strain peaks − Process cannot be applied to man-made fiber tapes and ropes	+ Silver-coated polymer thread: unsuitable as electrically conductive control tearing thread, since elongations at break cannot be reproduced or the parallel position of the untwisted filaments results in only individual filaments tearing in case of failure and the applied tension remaining constant − Metallic fine wire: very sensitive to breakage, otherwise excessively high elongation at break compared to load-bearing agent − Cellulose fiber with carbon filling: lower, moisture-dependent conductivity than silver-plated polyamide yarns or fine wires. − Optically conductive control thread: buckling sensitivity, critical mechanical behavior.
I MATERIAL PROPERTIES	Hardness: 50 ± 5 Shore A; density: 1.13 g/cm^3; tear strength: ≥3.5 N/mm^2; elongation at break: ≥200%; specific volume resistance: ≤12 Ωcm	
II ENERGY SUPPLY	Electric current	Electric current
III RESOLUTION		
IV SENSITIVITY		
V MEASUREMENT RANGE		
VI TRL	9	9

	INDUTECH	
	SENSOR FOR DETERMINING MANTLE SLIPPAGE	**FRICTION-SPUN ABRASION SENSOR THREAD**
SENSOR TYPE	Mechanical	Mechanical
MEASURAND	Electric current	Visual assessment
CONSTRUCTION PRINCIPLE	Thread, weave	Thread
GEOMETRY	Linear, planar	Linear
MATERIAL		Polypropylene, polyethylene terephthalate
Procedural principle	Sensor with optical signal output in the event of critical wear or damage to the outer sheath of a load-bearing rope or tape. The sensor thread has a core-sheath structure, the signal-colored core being sheathed with thermoplastic staple fiber. This has the color of the load-bearing textile and is integrated into its outer shell in such a way that it is exposed to abrasion during use. [4]	Sensor with optical signal output in the event of critical wear or damage to the outer sheath of a load-bearing rope or tape. The sensor thread has a core-sheath structure, the signal-colored core being sheathed with thermoplastic staple fiber. This has the color of the load-bearing textile and is integrated into its outer shell in such a way that it is exposed to abrasion during use. [4]
Schematic sketch	Electricity / Core / Mantle / Relative movement between mantle and core	Signal-colored core / Mantle of staple fibers in color of the load-bearing band
Known/possible field of application	Control and monitoring when guiding a rope with small deflection radii, since there can be relative movements between the core and sheath in the form of sheath slippage or compression and the introduction of forces into the textile sheath, which is designed as a non-load-bearing element, leads to impermissible wear on the rope.	Structural health monitoring of ropes.
Possible sensor variants	Arrangement of the conductor loops in dimensions that correspond to the pole configuration of a planar permanent magnet. Conductor loops as an execution of adjacent meshes through which a sectionally magnetized longitudinal structure can be moved. By matching conductor loops and permanent magnets, the induction voltages of all conductor loops add up.	Variation of the ratio of core to shell diameter. Core yarn made of PET, sheath yarn made of PP; core thread not signal-colored, but rather made of fluorescent material for UV detection. Variation of core and sheath strength. Core yarn made of PP, sheath yarn made of PET.
Opportunities and challenges	+ Accelerations due to relative movements of core and mantle are shown in a manner directly proportional to the magnitude of the stress induced in the meshes of the mantle + The magnetized threads can be oriented along the expected displacement and anchored to the core	+ The use of a fluorescing signal thread in the thread core enables an automated visual inspection of the wear condition by means of camera technology, even on soiled or very colorful load-bearing textiles + With increasing sheath fineness, there is a significant increase in bearable double chafing
MATERIAL PROPERTIES		
ENERGY SUPPLY	None	None
RESOLUTION		
SENSITIVITY		
MEASUREMENT RANGE		
TRL	9	6–8

	INDUTECH	
	WEAR SENSOR 2	**STRAIN SENSOR 2**
1 | SENSOR TYPE	Mechanical	Mechanical
2 | MEASURAND	Electric current	Electric current
3 | CONSTRUCTION PRINCIPLE	Thread	Thread
4 | GEOMETRY	Linear	Linear, diameter ø = 0.5–2.5 mm
5 | MATERIAL	Aramid	Carbon-black-filled silicone rubber
a) Procedural principle	Measurement of wear by a disruption sensor made of aramid. Carbon fibers integrated into a rope are contacted electronically. As the disintegration progresses, the electrical resistance increases, which serves as a characteristic value for wear.	Measurement of strains and strain peaks based on a reproducible dependence on strain and electrical resistance, even while maintaining the state of strain through plastic deformation.
b) Schematic sketch		
c) Known/possible field of application		Use for in situ monitoring and determination of load cycles.
d) Possible sensor variants		
e) Opportunities and challenges		+ For protection against overloading, it is not necessary for the sensor thread to fail. Exceeding a defined strain state is enough for the output of an alarm signal + By also detecting strain peaks, strain sensors open a wide range of applications, from crack sensors to sensors for detecting strain peaks
I MATERIAL PROPERTIES		Hardness: 50 ± 5 Shore A; density: 1.13 g/cm^3; tear strength: ≥3.5 N/mm^2; elongation at break: ≥200%; specific volume resistance: ≤12 Ωcm
II ENERGY SUPPLY	Electric current	Electric current
III RESOLUTION		
IV SENSITIVITY		
V MEASUREMENT RANGE		
VI TRL	6–8	9

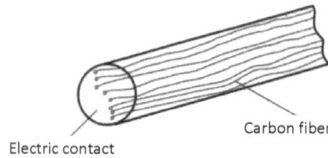

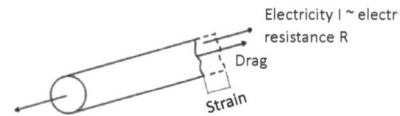

Schematic sketch (Wear Sensor 2): Electric contact — Carbon fiber

Schematic sketch (Strain Sensor 2): Electricity I ~ electr resistance R; Drag; Strain

SENSOR TYPE	Chemical
MEASURAND	Electromagnetic light spectrum
CONSTRUCTION PRINCIPLE	Weave
GEOMETRY	Planar
MATERIAL	FBG fibers in fabric
Procedural principle	FBG is a distributed Bragg reflector constructed in a short segment of optical fiber that reflects wavelengths of light and transmits all others. This is achieved by creating a periodic variation in the refractive index of the fiber core, which generates a wavelength-specific dielectric mirror. [23]
Schematic sketch	
Known/possible field of application	Used in seismology, pressure sensors for extremely harsh environments, and downhole sensors in oil and gas wells for measurement of the effects of external pressure, temperature, seismic vibrations and inline flow measurement.
Possible sensor variants	Integration of Bragg fiber as warp thread; into a 3D woven; embedded in a conveyor belt; inserted into a groove and threaded into flat-woven fabric.
Opportunities and challenges	+ Inline optical filter to block certain wavelengths, or as a wavelength-specific reflector
MATERIAL PROPERTIES	
ENERGY SUPPLY	Light
RESOLUTION	
SENSITIVITY	
MEASUREMENT RANGE	
TRL	<6

MEDTECH

	INTEGRATED OPTICAL FREQUENCY DOUBLER	TEMPERATURE SENSOR
1 | SENSOR TYPE	Thermal	Thermal
2 | MEASURAND	Electromagnetic light spectrum	Electric current, electromagnetic light spectrum, transmitted light, temperature
3 | CONSTRUCTION PRINCIPLE	Fiber-optic conductor	Thread
4 | GEOMETRY	Linear	Linear
5 | MATERIAL		Metals, electrically conductive polymers, glass fibe
a) Procedural principle	Determination of absolute temperatures by means of optical frequency doubling, in which a special light wavelength is required for a known temperature of the resonator in order to achieve a frequency conversion (phase matching of fundamental and harmonic wave) with high efficiency. [1]	Design of thread-shaped sensors for the investigation of thermal loads based on low-meltir metal wires, which change their electrical properti(under thermal load. [4] Temperature determination by measuring the change of the refraction coefficient of the light-gui(sheath under temperature change, which leads to corresponding transmission difference. [9]
b) Schematic sketch	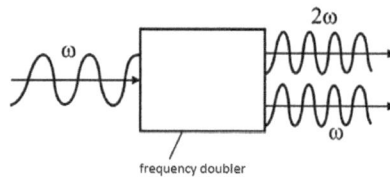	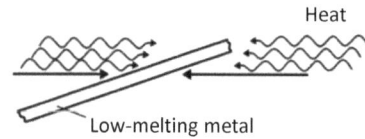
c) Known/possible field of application	Temperature monitoring of textile structures.	Temperature monitoring of textile structures.
d) Possible sensor variants	Particularly high efficiency.	Use of threads of electrically conductive polymers or electrically conductive coated polymers. Temperature sensors based on the principle of absorption edge displacement using filter glasses instead of semiconductor elements.
e) Opportunities and challenges	− The prerequisite for measurement is a tunable, coherent light source with enough power to operate the resonator	+ High reproducibility + Short response time + High accuracy + Low tendency for thread or surface production due to unfavorable properties of the metals
I MATERIAL PROPERTIES		
II ENERGY SUPPLY	Electric current	None
III RESOLUTION		
IV MATERIAL		
V MATERIAL PROPERTIES		50–250 °C
VI TRL	6–8	6–8

Schematic sketch labels: 2ω, ω, frequency doubler; Heat, Low-melting metal

	MEDTECH	
	RAPID-SHRINK FIBER	**STRAIN/PRESSURE SENSOR**
SENSOR TYPE	Chemical	Mechanical
MEASURAND	Visual assessment	Electric current
CONSTRUCTION PRINCIPLE	Warp knit, weave	Weft knit, warp knit
GEOMETRY	Linear, planar	Planar
MATERIAL	Elastomer (polyutherane, rubber)	Stainless steel
Procedural principle	A polymer fiber which shrinks rapidly at ordinary temperature and in contact with water, but retains its shape (impact strength), has high absorbency and has performance characteristics such as rubber elasticity. [14]	Spacer weft-knit made of electrically conductive stainless-steel-fiber yarns for detecting the position of the contact and the size of the contacting surface when the specific electrical resistance of the electrically conductive conductor paths changes as a result of elongation or pressure. [21]
Schematic sketch		
Known/possible field of application	Disposable diapers; fastening tapes; cloths as covers for dampening units in offset printers; cords or cylinders for plant cultivation; cords and nets for the food industry; bank reinforcements.	Flat pressure load in buildings.
Possible sensor variants	A water-absorbing, shrinkable yarn produced by blending or by blending spinning the rapidly shrinking fiber and a fiber that shrinks slower than said fiber upon absorption of water. A water-absorbing shrinkable material which consists of a water-absorbing shrinkable fibrous web and a water-absorbing shrinkable yarn that absorbs water at a higher rate and to a greater extent than the fibrous web, with the water-absorbing shrinkable yarn containing the rapidly shrinking fiber.	Spacer warp-knit can also be used instead of spacer weft-knit.
Opportunities and challenges		– Spacer warp-knit has a hysteretic force behavior and is therefore less suitable as a pressure sensor
MATERIAL PROPERTIES	At 20 °C, maximum percentage shrinkage >30%	
ENERGY SUPPLY		Electric current
RESOLUTION	0–10 s	
MATERIAL		
MATERIAL PROPERTIES	At 20 °C, shrinkage stress = 0.351–1.755 kg/m² (30–150 mg/den)	
TRL	9	6–8

Schematic sketch labels (Rapid-shrink fiber): Radius r; Water-absorbent shrinkable yarn; Rapidly shrinking fiber; Pitch d; Twist angle

Schematic sketch labels (Strain/pressure sensor): Knitted surface; Y; Z; X; D; L; B; Polfiber

	TEXTILES WITH SPECIAL FUNCTIONS	CLOTHING INDICATOR FOR UV RADIATION AND OZONE
1 | SENSOR TYPE	Chemical	Chemical
2 | MEASURAND	Visual assessment, electric current	Visual assessment
3 | CONSTRUCTION PRINCIPLE		
4 | GEOMETRY		
5 | MATERIAL	Hollow polymers modified with moisture-sensitive gels	
a) Procedural principle	Garment comprising sensory and/or actuatorically modified polymers which, in the event of a health and/or environmental hazard, change their color, geometric shape or other physical, biological or chemical properties to protect the wearer in a defined manner. [27]	Small-scale application of a variety of fashionable design forms whose color change is accompanied by influencing factors from the environment and at least semi-quantitatively correlates with the hazard potential. The measurement is carried out either by the iodine method, acetone decomposition, oxalic acid decomposition or an immune globuline (IG) dosimeter.
b) Schematic sketch		

Basic thread in right-left-binding

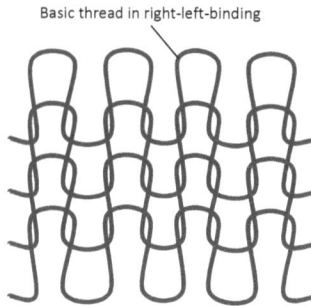

c) Known/possible field of application	Monitoring of danger conditions.	Sensor application to swimwear, leisurewear and workwear for outdoor activities for the detection of UV radiation.
d) Possible sensor variants	Sensor element can be formed from: temperature sensors, pressure sensors, humidity sensors, pH value sensors, radiation sensor.	Determination of the intensity of UV radiation by measurement using iodine method, acetone decay, oxalic acid decomposition or IG dosimeter.
e) Opportunities and challenges		+ Good resistance of the textile carrier to the hazard potential, sensitization technology and reactions causing color change
		− Doubts as to whether the concentration and intensity of the hazard potential is sufficient to initiate the chemical reaction on the textile
I MATERIAL PROPERTIES		
II ENERGY SUPPLY		Light
III RESOLUTION		
IV SENSITIVITY		
V MEASUREMENT RANGE		
VI TRL	6–8	<6

MEDTECH

	PHYSIOLOGICAL SENSOR 1	PHYSIOLOGICAL SENSOR 3
SENSOR TYPE	Chemical, mechanical	Thermal
MEASURAND	Electric current	Electromagnetic light spectrum
CONSTRUCTION PRINCIPLE	Conductive yarn	Weft knit, warp knit, weave
GEOMETRY	Planar	Sheath diameter: 0.125 mm; core diameter: 0.09 mm
MATERIAL	Silver/gold-coated nylon	
Procedural principle	Recording of physiological states via sensors, transmission via electrically conductive conductor paths in clothing and processing in measuring equipment. [31]	Integration of a fiber-optic temperature-sensing element into a fabric The temperature sensing element is an optical fiber containing one or more fiber-Bragg-grating sensors. Light is introduced into the optical single-mode fiber and directed to a grating interface adjacent to the wearer. A reflux signal is received by a reflection mode or a transmission mode, the reflux signal having a wavelength shift indicative of temperature by the Bragg resonance effect. [42]
Schematic sketch		

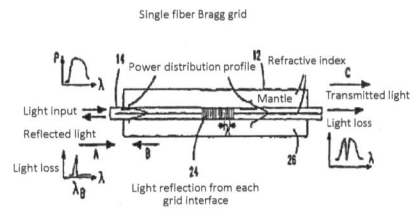

Single fiber Bragg grid

	PHYSIOLOGICAL SENSOR 1	PHYSIOLOGICAL SENSOR 3
Known/possible field of application	Sportswear and medical clothing for monitoring bodily functions. Multimedia clothing for adapting media enjoyment to physiological conditions.	
Possible sensor variants	Sportswear/medical clothing. [32], [33], [34], [35], [36] Textile electrode in spacer warp-knit. [37], [38], [39] Multifunctional apparel system. [40]	Processing of the thread in a weft knit, warp knit, or weave.
Opportunities and challenges	+ Advantageous contact behavior due to pressure-elastic behavior when using monofilaments + Acceptance by the wearer due to attractive appearance + Comfortable to wear due to the flexibility of the garment	
MATERIAL PROPERTIES	Electrical resistance: <5 Ω/cm; diameter of monofilaments: >100 µm	
ENERGY SUPPLY	Electric current	Electric current
RESOLUTION		
MATERIAL		
MATERIAL PROPERTIES		
TRL	9	6–8

MEDTECH

	PHYSIOLOGICAL SENSOR 4	PHYSIOLOGICAL SENSOR 5
1 | SENSOR TYPE	Mechanical, chemical, thermal	Mechanical
2 | MEASURAND	Electric current	Electric current
3 | CONSTRUCTION PRINCIPLE	Elastic weave or fleece containing electrically conductive fibers	Weft knit containing electrically conductive thread
4 | GEOMETRY	Linear, planar	Punctiform, linear or planar
5 | MATERIAL	Elastomers filled with conductive particles or electrically conductive metals	
a) Procedural principle	Garment with belts running transversely to the longitudinal axis of the wearer, which can be stretched in the longitudinal direction and in which strain gauges are incorporated, which allow physiological functions to be determined via changes in electrical conductivity. [43]	Sensor consisting of strain gauges, piezoelectric elements, length gauges or pressure sensors, all of which change their electrical properties under mechanical deformation. [44]
b) Schematic sketch		

Conductor track — Connection wire — Connection wire — Conductor track — Sensor — Belt — Back section — Leg

	PHYSIOLOGICAL SENSOR 4	PHYSIOLOGICAL SENSOR 5
c) Known/possible field of application	Clothing for monitoring heart activity and recording skin resistance, perspiration and body temperature.	Garment for determining a posture or movement of the body.
d) Possible sensor variants	The carrier material of the electrically conductive threads is knitted fabric made of cotton with elastane content or viscose, or synthetic or microfiber. Conductive particles in the elastor of the strain sensor can be carbon particles or hydrogels.	Sensor element can be formed from strain gauges.
e) Opportunities and challenges	+ The garment should be resistant to perspiration and to washing + Increase of sensor sensitivity through path-shaped guidance of the strain sensor, since the transverse elongation is low compared to a longitudinal elongation − An insulating layer should prevent moisture from influencing the measuring signal of the extensometers − The elastomer should be more extensible than the substrate on which the sensor is placed so that the extensibility of the sensor does not limit that of the garment	+ Piezoelectric elements + Magnetic, capacitive or optical length gauges. + High wearing comfort due to the unobtrusive integration of the sensor elements into the garment
I MATERIAL PROPERTIES	Specific sensor resistance: 5–30,000 Ωcm	
II ENERGY SUPPLY	Electric current	Electric current
III RESOLUTION		
IV SENSITIVITY		
V MEASUREMENT RANGE		
VI TRL	9	9

MEDTECH

	TENSILE STRAND	PHYSIOLOGICAL STRAIN SENSOR
SENSOR TYPE	Mechanical	Mechanical
MEASURAND	Electric current	Electric current
CONSTRUCTION PRINCIPLE	Weft knit, warp knit, weave	Elastic fabric containing electrically conductive filaments
GEOMETRY	Linear, planar	Linear, planar
MATERIAL	Elastic, electrically conductive core thread; bimetallic sheathing	
Procedural principle	Thread for determining the tensile stress. The thread consists of an elastic core thread and has at least one sheath, this sheath changing its electrical property, in particular its electrical resistance and/or its capacitance, when the length of the core thread changes. As a result of the tensile stress, the pressure is exerted on the body of the wearer. [45]	Electrically conductive thread for determining the state of respiration and movement, which changes its electrical properties under tensile or compressive load, above all its electrical resistance and inductance. [46]
Schematic sketch	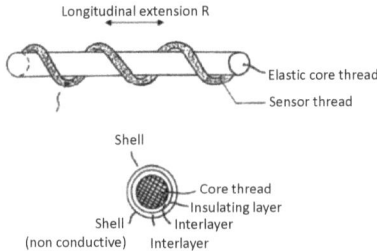	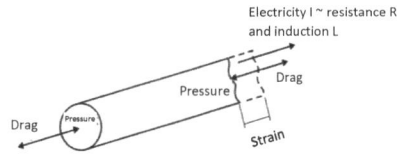
Known/possible field of application	Bandage or compression stocking. Sheathing can release substances to the skin of the wearer.	Clothing for monitoring respiratory and physical activity of newborns, children, adults and even non-human mammals.
Possible sensor variants	Processing of the thread—preferably as weft thread—in a weft knit, warp knit, or weave.	Clothing for monitoring respiratory and physical activity of newborns, children, adults and even non-human mammals.
Opportunities and challenges	+ Single- or multilayer sheathing + Thread with one or more wrapping threads	+ Moisture resistant, i.e., washable + High measurement accuracy + No slipping of the sensors, due to precise positioning in the garment + Increased wearing comfort due to the nottoo tight fit of the garment
MATERIAL PROPERTIES		
ENERGY SUPPLY	Electric current	Electric current
RESOLUTION		
SENSITIVITY		
MEASUREMENT RANGE		
TRL	9	9

	PRESSURE SENSOR	KNITTED BREATHING SENSOR
1 | SENSOR TYPE	Mechanical	Mechanical
2 | MEASURAND	Electric current, pressure	Electric current
3 | CONSTRUCTION PRINCIPLE	Elastic weft-knit or warp-knit containing electrically conductive threads	
4 | GEOMETRY	Dimensions of the pressure sensor: 10 mm x 10 mm x 1 mm	
5 | MATERIAL	Electrically conductive metals	Electrically conductive polyester with stainless-stee content
a) Procedural principle	Pressure sensor with strip-like or filament-like elements which each have a layered structure and are electrically conductive. When pressure is applied, the layers touch each other and a closed circuit is formed which indicates the pressure. [47]	Measurement of respiratory movement by changin the electrical resistance when weft knits made of conductive polyester fiber yarns with stainless-stee content are stretched. [48]
b) Schematic sketch		
c) Known/possible field of application	Clothing for monitoring heart activity and recording skin resistance, perspiration and body temperature.	
d) Possible sensor variants	Pressure sensitive stocking. [49]	Right/left weft knit with conductive stripes. Right/left lining weft knit in which the conductive yarn no longer forms any stitches, but is merely integrated with handles in the non-conductive basi knit Right/right weft knits where the electrical resistanc is less dependent on elongation
e) Opportunities and challenges		
I MATERIAL PROPERTIES	Environmentally stable at 0–50 °C, 30–90% relative humidity	
II ENERGY SUPPLY	Electric current	Electric current
III RESOLUTION		
IV SENSITIVITY		
V MEASUREMENT RANGE	Surface pressure: 0–10 kg/cm²	
VI TRL	6–8	6–8

	MEDTECH	
	THREE-DIMENSIONAL SPACER WARP-KNIT	**FIBER-OPTIC RESONATOR**
SENSOR TYPE	Mechanical	Thermal
MEASURAND	Electric current	Electromagnetic light spectrum, temperature
CONSTRUCTION PRINCIPLE	Warp knit	Fiber-optic conductor
GEOMETRY	Planar	Linear
MATERIAL		
Procedural principle	Three-dimensional spacer warp-knit with integrated ultrasonic sensors for monitoring body movement. [50]	Stabilization of the system consisting of a semiconductor laser and a fiber-optic resonator in a resonance, i.e., at maximum transmission and minimum reflection, respectively, by retuning the wavelength of the laser light by the operating current when the temperature and the optical path length of the fiber-optic resonator change. [1]
Schematic sketch	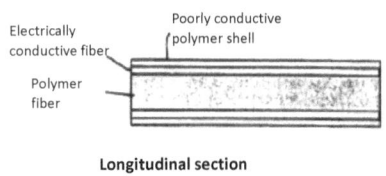	
Known/possible field of application		Medical applications.
Possible sensor variants	Flexible elastane material guarantees flexibility and wearing comfort.	Possibility to measure absolute temperatures after calibration of the system.
Opportunities and challenges		− Restricted measuring range
MATERIAL PROPERTIES		
ENERGY SUPPLY	Electric current	Electric current
RESOLUTION		
SENSITIVITY		
MEASUREMENT RANGE		20 °C
TRL	6–8	6–8

Electrically conductive fiber — Poorly conductive polymer shell — Polymer fiber

Longitudinal section

MEDTECH

	MINIATURE SENSOR FOR CHECKING SEAM AND THREAD TENSION	GLASS FIBER SENSOR
1 | SENSOR TYPE	Mechanical	Chemical
2 | MEASURAND	Electric current, electromagnetic light spectrum	Electromagnetic light spectrum
3 | CONSTRUCTION PRINCIPLE	Yarn	Fiber
4 | GEOMETRY	Linear	Linear
5 | MATERIAL		Glass
a) Procedural principle	Microsensors for detection of elongation, bending, displacement or pressure in thread and seam. When the thread is subjected to tensile stress, there is a characteristic change in the electrical properties (electrical resistance, capacitance, inductance, electromagnetism) or in the transmission behavior of the light in the thread. [58]	A change in the light frequency due to the measured variable to be determined can be recorded as a typical characteristic value. [59]
b) Schematic sketch	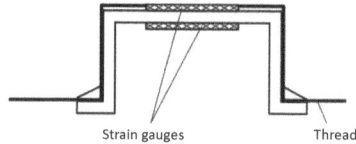	
c) Known/possible field of application	Medical applications (measurement of seam and thread tension to avoid excessive thread tension).	Fiber-optic sensors in chemical process control, automotive engineering, shipbuilding, mining, medicine and nuclear industry. Level sensor.
d) Possible sensor variants	Measurement via the electrical resistance: strain gauge, direct or hydrostatic, coupled pressure sensor, linear potentiometer. Measurement over capacity: deformed dielectric, variable pitch capacitor, plate spacing. Measurement via inductance: coil with movablecore, differential transformer. Measurement via electromagnetism: force–current converter. Measurement via magnetism: reverberation effect. Measurement via light: curved/drawn fiber, light-emitting diode and quadrant diode.	Sensor system differentiated according to fiber-optic transmitter, hybrid system and use of the fiber itself as a sensor (internal/external modulation). Alternative use of plastic fibers in short-distance systems.
e) Opportunities and challenges	+ No available data concerning the biological tissue, since it is generally inhomogeneous, anisotropic and highly time-variant; inhomogeneity of the tissue makes precise measurement necessary; compatibility of the sensor in the organism; limited installation space; high sensitivity necessary for low measuring range	+ Low volume + Low weight + Galvanic isolation of input and output makes earthing unnecessary + No interference from external electromagnetic fields + No danger of explosion − Problems with the coupling of fibers
I MATERIAL PROPERTIES		
II ENERGY SUPPLY	Electric current	Light
III RESOLUTION		
IV SENSITIVITY		
V MEASUREMENT RANGE		
VI TRL	<6	9

Caption for schematic sketch: Strain gauges   Thread

	MEDTECH	
	3D TUBULAR FABRIC	**CAPACITIVE BREATHING SENSOR**
SENSOR TYPE	Mechanical	Mechanical
MEASURAND	Electric current	Electromagnetic field
CONSTRUCTION PRINCIPLE	Tube fabric	Conductive ink between textile layers
GEOMETRY	Three-dimensional	Planar
MATERIAL	Polyester-laminated aluminum tape fabric	Combination of stretchable and non-stretchable textile with conductive ink
Procedural principle	Tubular fabric in which conductive aluminum ribbons are woven. Under pressure load, the hose is compressed and acts as a condenser. This produces a voltage change that can be correlated with the pressure load.	Respiration measurement via capacitive proximity sensor. Respiratory frequency is measured by the displacement of two textile layers, which are connected by a conductive layer, caused by the respiratory movement. [54]
Schematic sketch		
Known/possible field of application	Measuring changes in pressure load, e.g., decubitus prohylaxis or fall prevention; improvement of ergonomics.	
Possible sensor variants	Temporal resolution subject to sensor design.	
Opportunities and challenges	+ Tubular shape allows even compression under strain without the risk of the conductive belts shifting against each other + Production in one weaving process possible − Correlation of position/load and measuring signal for each position of the ligament tissue to be re-determined	
MATERIAL PROPERTIES	Operating range -20 to +50 °C	
ENERGY SUPPLY	Electric current	Electric current
RESOLUTION	250 ms	
SENSITIVITY	Changes in capacitance of 2%	
MEASUREMENT RANGE	Pressure up to 150 kg	
TRL	6–8	<6

	CARBON NANOTUBE (CNT) STRAIN SENSOR	PRESSURE MAPPING SYSTEM
1 | SENSOR TYPE	Mechanical	Mechanical
2 | MEASURAND	Electric current	Electric current
3 | CONSTRUCTION PRINCIPLE	Yarn	Three layers of fabric
4 | GEOMETRY	Linear	Planar
5 | MATERIAL	Carbon nanotube yarn	Piezoresistive semiconductive polymers between two layers of highly conductive ripstop nylon fabric
a) Procedural principle	Electrical resistance of twisted CNT yarns changes with load or temperature change. [55]	Using piezoresistive sensors to quantify the pressu between two contacting objects, such as a person and his or her support surface. [23]
b) Schematic sketch		
c) Known/possible field of application	Monitoring of motion and temperature.	Monitoring of vital functions.
d) Possible sensor variants	An advanced strain sensor for human motion detection was introduced by Yamada. It uses a new material, namely thin films of aligned single-walled carbon nanotubes. Unlike traditional rigid materials such as silicon, nanotube films fracture into gaps and islands, and bundles bridge the gaps. This allows the films to function as strain sensors capable of measuring strains of up to 280% with high durability.	Thin mats are composed of a matrix of small senso and a cover. When a person sits on such a mat, the sensors read pressure at individual locations on the thigh or buttock. This data is transferred to a computer, where a clinician can analyze it. Evenly distributed pressure is preferred. Used by clinicians to determine the suitability of a wheelchair cushion, and by researchers investigating support surfaces, risk factors for ulceration and ulcer prevention protocols. Used in industrial and engineering environments fo product design and verification, process control or quality assurance.
e) Opportunities and challenges		
I MATERIAL PROPERTIES		
II ENERGY SUPPLY	Electric current	Electric current
III RESOLUTION		
IV SENSITIVITY	Strain: 1.4 to 1.8 mV/V/1000 m; temperature: 91 mA/°C	
V MEASUREMENT RANGE		
VI TRL	<6	9

	MEDTECH	
	CAPACITIVE PRESSURE SENSOR	**BIOPOTENTIAL SENSORS**
SENSOR TYPE	Mechanical	Mechanical
MEASURAND	Electric current	Electric current
CONSTRUCTION PRINCIPLE	Weave	Weaves, weft knits, embroidered electrodes
GEOMETRY	Planar	Planar
MATERIAL	Piezoresistive material between fabric layers	Silver yarns for electrodes
Procedural principle	Resistive pressure sensor is comprised of a matrix of capacitive-sensing elements. Pressure applied to the surface of the sensing element causes a change in capacitance that is correlated to a change in pressure. Proprietary Windows-based software compensates for sensor non-linearity, hysteresis and creep over time, resulting in enhanced accuracy. [23]	Electrocardiography (ECG) and electromyography (EMG) are the electrical potentials periodically changed by cardiovascular and muscular activities. [60]
Schematic sketch		
Known/possible field of application	Monitoring of vital functions.	
Possible sensor variants	Capacitive-based pressure-imaging sensors developed by XSENSOR Technology Corporation (Calgary, Alberta, Canada) can graphically display pressure distributions in real time between virtually any two surfaces in contact. The sensor element is accurate, thin, flexible and robust. These physical characteristics minimize any artificial influences created by the presence of the sensor during data collection.	Nervous stimuli and muscle contraction can be easily detected by measuring the ionic current flow in the body. This measurement is accomplished by attaching biopotential electrodes to the skin surface. ECG/EMG-monitoring systems: the electrodes are either made of gel or stuck to the skin using conductive adhesives in order to develop better contact to the skin. To improve contact between the electrodes and the skin, skin preparation is required, such as shaving, abrading, and cleaning the skin surface. Wearable electrode is created by weaving, knitting or stitching silver yarns on the inner surface of the clothing. Irregular surface structures create high impedance, and therefore high-frequency noise.
Opportunities and challenges	+ Accurate, thin, flexible and robust	− Gelatinous substances dry out over a long period of time and cause the electrode to come off the skin. Adhesives can irritate the skin, leading to a loss of signal quality
MATERIAL PROPERTIES		
ENERGY SUPPLY	Electric current	Electric current
RESOLUTION		
SENSITIVITY		
MEASUREMENT RANGE		
TRL	9	9

	MEDTECH	
	RESPIRATORY SENSORS INDUCTIVE PLETHYSMOGRAPHY (RIP)	**FLEXIBLE SKIN-ATTACHABLE PIEZOELECTRIC SENSOR**
1 | SENSOR TYPE	Mechanical	Mechanical
2 | MEASURAND	Electric current	Electric current
3 | CONSTRUCTION PRINCIPLE	Metal wire in textile strips	Scrim
4 | GEOMETRY	Linear	Planar
5 | MATERIAL	Metal wires	PVDF TrFE nanofiber material and Au-sputtered PDMS sheets
a) Procedural principle	RIP signals can be caught by an insulated sinusoidal wire coil embedded into a stretchable textile strap. [60]	Piezoelectric nanofiber-based sensors made from electrospun nanofiber material of poly(vinylidenefluoride-co-trifluoroethylene) (PVD TrFE) that is sandwiched between two elastomer sheets with gold-sputtered electrodes as an active layer. [54]
b) Schematic sketch		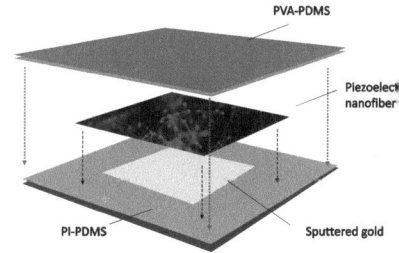
c) Known/possible field of application		
d) Possible sensor variants	Wound around the chest or abdomen, the textile strap is intended to be stretched by respiration. The coil inductance is directly governed by the change of sinusoid shapes.	Targeted uses as a high-precision pulse-monitoring device.
e) Opportunities and challenges		+ Ultra-thin, stretchable, flexible sensor that can b attached to the skin
I MATERIAL PROPERTIES		
II ENERGY SUPPLY	Electric current	Electric current
III RESOLUTION		10 µm skin displacement
IV SENSITIVITY		1 µm coefficient of variation
V MEASUREMENT RANGE		
VI TRL	9	6–8

Schematic sketch labels: PVA-PDMS, Piezoelectric nanofiber, PI-PDMS, Sputtered gold

MEDTECH

	PRESSURE FORCE MAPPING SENSOR	TEXTILE-BASED GONIOMETER
SENSOR TYPE	Mechanical	Mechanical
MEASURAND	Electric current	Electric current
CONSTRUCTION PRINCIPLE	Weave, weft knit	Weft knit
GEOMETRY	Planar	Planar
MATERIAL	Carbon black, metal, and metal oxide particles	Combination of electroconductive (Belltron® by Kanebo Ltd.) and elastic (Lycra®) yarns
Procedural principle	The technology behind force mapping is typically a grid of individual force sensor elements. The core principle of electrical resistance-based pressure mapping is the special property of electrically conducting polymer composites (ECPC), that their deformation, which could be caused by either tension or pressure, will cause its electrical impedance in the vicinity of the deformation to change. [56]	Knitted piezoresistive fabrics modify their electrical resistance when they are elongated or flexed. The main requirement for the application of the single-layer sensors is that the human movements must produce a strain field which can be detected in terms of resistance variation. For this reason, single-layer sensors must be integrated into adherent garments close to the human joint under investigation. [25]
Schematic sketch	Layer stacks	
Known/possible field of application		Hand motion sensing: a kinesthetic sensing glove was developed for the ambulatory evaluation of the residual hand function and its recovery in post-stroke patients; scapular movement detection.
Possible sensor variants	Force sensors can be implemented based on various principles, such as piezoresistive, piezoelectric, piezomagnetic, capacitive, magnetic and optical. The basic physical structure of capacitive-based pressure mapping sensors is two parallel conductive plates separated with a flexible, non-conductive layer as the dielectric spacer.	Single-layer sensor or double-layer sensors.
Opportunities and challenges	+ The sensing elements can be isolated from the skin by either additional regular textile layers or direct isolation coatings to avoid any complications from electrode–skin contact. + Easily scalable in terms of sensing channels; this is mainly because of the simplicity of the measuring structure − Higher data processing/transmission requirements; the need for special conductive and/or dielectric materials; relatively complex sensor structures	− Needs to closely adhere to joint moving
MATERIAL PROPERTIES		
ENERGY SUPPLY	Electric current	Electric current
RESOLUTION	40 Hz	
SENSITIVITY		Single layer: 6405 Ω for $\Delta\theta = 37°$
MEASUREMENT RANGE		Double layer: 5100 Ω for $\Delta\theta = 37°$
TRL	6–8	6–8

HOMETECH

1 | SENSOR TYPE	Mechanical, chemical
2 | MEASURAND	Visual assessment
3 | CONSTRUCTION PRINCIPLE	n/a
4 | GEOMETRY	Linear
5 | MATERIAL	n/a

a) Procedural principle

Permanent identification of harmful environmental influences through the use of threads which change their shape, color or volume while absorbing liquids. The core yarn must be UV-resistant and clearly distinguished in color from the load-bearing tape. For the sheath fibers of the yarn, a material must be selected which is changed in shape, color or structure by UV radiation. [4]

b) Schematic sketch

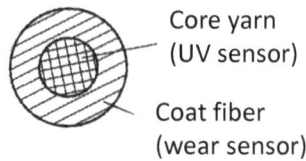

Core yarn
(UV sensor)

Coat fiber
(wear sensor)

c) Known/possible field of application

A friction-spun sensor thread represents a combination of an abrasion sensor and a UV sensor.

d) Possible sensor variants

Decrease in abrasion resistance with increasing exposure to UV radiation.

e) Opportunities and challenges

I MATERIAL PROPERTIES	
II ENERGY SUPPLY	None
III RESOLUTION	
IV SENSITIVITY	
V MEASUREMENT RANGE	
VI TRL	9

	HOMETECH	
	TEMPERATURE-CONTROLLED RADIATION TRANSMISSION	**INTELLIGENT MEMBRANE**
SENSOR TYPE	Mechanical	Mechanical
MEASURAND	Electric current	Electromagnetic light spectrum, electric current, noise level
CONSTRUCTION PRINCIPLE	Fleece	Fiber bundle in woven or knitted structure
GEOMETRY	Flat, fiber diameter of 0.01 to 10 mm	Areal, change of shape up to 8 times its size
MATERIAL	Thermotropic polymer blends	Nickel titanium alloy
Procedural principle	A polymer-based material having temperature-controlled radiation transmission which is present within core/sheath fibers in a core. A transparent shell surrounds the core of thermotropic polymer mixture, which becomes turbid beyond the so-called lower critical demixing temperature (LCST) due to a changing radiation emission. This turbidity effect occurs due to a structural change in the polymer system, in which the components with different refractive indices separate due to temperature change. A variation of the relative contents of the individual comonomers causes turbidity at different temperatures. [17]	A membrane with built-in sensors which reacts to stimuli such as light, contact, noise or environmental movements in a mobile manner via muscle wires made of Ni-Ti alloy developing different temperatures at certain currents and passing through different movements. [22]
Schematic sketch		

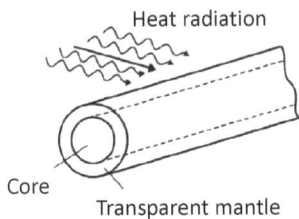

Heat radiation
Core
Transparent mantle

Known/possible field of application	Temperature-dependent control of radiation transmission on buildings (awnings, roller blinds, venetian blinds), technical equipment, in the clothing industry and for decorative purposes.	
Possible sensor variants	Incorporation of a non-thermotropic but mechanically highly resilient material into the polymer core.	Use in any size possible.
Opportunities and challenges	+ Advantage of core-shell structure when using aids with low compatibility to thermotropic core material − Expensive production − Bonding of polymers only possible at high application temperatures − Limited possibility of reversible structural change − Low mechanical load capacity	− Very expensive materials
MATERIAL PROPERTIES	Relative proportion of comonmers between 0.1 and 50 mol%	
ENERGY SUPPLY	Electromagnetically	Electric current
RESOLUTION		
SENSITIVITY		
MEASUREMENT RANGE		
TRL	9	<6

	HOMETECH	
	ALARM WALLPAPER	**3D TUBULAR FABRIC**
1 | SENSOR TYPE	Mechanical	Mechanical
2 | MEASURAND	Electric current	Electric current
3 | CONSTRUCTION PRINCIPLE	Fiber fleece	Tube fabric
4 | GEOMETRY	Planar	Three-dimensional
5 | MATERIAL	Fiber fleece: plastic; conductor paths: electrically conductive metals	Polyester-laminated aluminum tape fabric
a) Procedural principle	Surface monitoring system with coating of plastic fiber fleece coated with electrically conductive, metal-free conductive tracks which trigger an alarm in case of damage. [15]	Tubular fabric in which conductive aluminum ribbons are woven. Under pressure load, the hose compressed and acts as a condenser. This produce a voltage change that can be correlated with the pressure load.
b) Schematic sketch		
c) Known/possible field of application	Alarm in case of damage to surfaces.	Measuring changes in pressure load, e.g., decubitu prohylaxis or fall prevention; improvement of ergonomics.
d) Possible sensor variants	Simple retrofitting possible.	Temporal resolution subject to sensor design.
e) Opportunities and challenges	+ Self-calibration function + Device not detectable via instruments + Side-effects (noises, vibrations and temperature fluctuations) are not recorded + Modular system structure possible + Roll material for use in all cases of need − Impairment by nails, screws or dowels in the wall	+ Tubular shape allows even compression under strain without the risk of the conductive belts shifting against each other + Production in one weaving process possible − Correlation of position/load and measuring signal for each position of the ligament tissue to be re-determined
I MATERIAL PROPERTIES		Operating range -20 to +50 °C
II ENERGY SUPPLY	Electric current	Electric current
III RESOLUTION		250 ms
IV SENSITIVITY		Changes in capacitance of 2%
V MEASUREMENT RANGE		Pressure up to 150 kg
VI TRL	6–8	6–8

	HOMETECH	
	PRESSURE MAPPING SYSTEM	**CAPACITIVE PRESSURE SENSOR**
SENSOR TYPE	Mechanical	Mechanical
MEASURAND	Electric current	Electric current
CONSTRUCTION PRINCIPLE	Three layers of fabric	Weave
GEOMETRY	Planar	Planar
MATERIAL	Piezoresistive semiconductive polymers between two layers of highly conductive ripstop nylon fabric	Piezoresistive material between fabric layers
Procedural principle	Using piezoresistive sensors to quantify the pressure between two contacting objects, such as a person and his or her support surface. [23]	The resistive pressure sensor is comprised of a matrix of capacitive-sensing elements. Pressure applied to the surface of the sensing element causes a change in capacitance that is correlated to a change in pressure. Proprietary Windows-based software compensates for sensor non-linearity, hysteresis and creep over time, resulting in enhanced accuracy. [23]
Schematic sketch		
Known/possible field of application	Monitoring of vital functions.	Monitoring of vital functions.
Possible sensor variants	Thin mats are composed of a matrix of small sensors and a cover. When a person sits on such a mat, the sensors read pressure at individual locations on the thigh or buttock. This data is transferred to a computer, where a clinician can analyze it. Evenly distributed pressure is preferred. Used by clinicians to determine the suitability of a wheelchair cushion, and by researchers investigating support surfaces, risk factors for ulceration and ulcer prevention protocols. Used in industrial and engineering environments for product design and verification, process control or quality assurance.	Capacitive-based pressure-imaging sensors developed by XSENSOR Technology Corporation (Calgary, Alberta, Canada) can graphically display pressure distributions in real time between virtually any two surfaces in contact. The sensor element is accurate, thin, flexible and robust. These physical characteristics minimize any artificial influences created by the presence of the sensor during data collection.
Opportunities and challenges		+ Accurate, thin, flexible and robust. Accurate, thin, flexible and robust
MATERIAL PROPERTIES		
ENERGY SUPPLY	Electric current	Electric current
RESOLUTION		
SENSITIVITY		
MEASUREMENT RANGE		
TRL	9	9

	PRESSURE FORCE MAPPING SENSOR	TEXTILE-BASED GONIOMETER (TBC)
1 | SENSOR TYPE	Mechanical	Mechanical
2 | MEASURAND	Electric current	Electric current
3 | CONSTRUCTION PRINCIPLE	Weave, weft knit	Weft knit
4 | GEOMETRY	Planar	Planar
5 | MATERIAL	Carbon black, metal, and metal oxide particles	Combination of electroconductive (Belltron® by Kanebo Ltd.) and elastic (Lycra®) yarns
a) Procedural principle	The technology behind force mapping is typically a grid of individual force sensor elements. The core principle of electrical resistance-based pressure mapping is the special property of electrically conducting polymer composites (ECPC), that their deformation, which could be caused by either tension or pressure, will cause their electrical impedance in the vicinity of the deformation to change. [56]	Knitted piezoresistive fabrics modify their electrical resistance when they are elongated or flexed. The main requirement for the application of the single-layer sensors is that the human movements must produce a strain field which can be detected in terms of resistance variation. For this reason, single-layer sensors must be integrated into adherent garments close to the human joint under investigation. [25]
b) Schematic sketch	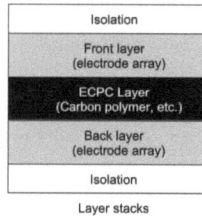 Layer stacks	
c) Known/possible field of application		Hand motion sensing: a kinesthetic sensing glove was developed for the ambulatory evaluation of the residual hand function and its recovery in post-stroke patients; scapular movement detection.
d) Possible sensor variants	Force sensors can be implemented based on various principles such as piezoresistive, piezoelectric, piezomagnetic, capacitive, magnetic and optical. The basic physical structure of capacitive-based pressure-mapping sensors is two parallel conductive plates separated with a flexible, non-conductive layer as the dielectric spacer.	Single-layer sensor or double-layer sensors.
e) Opportunities and challenges	+ The sensing elements can be isolated from the skin by either additional regular textile layers or direct isolation coatings to avoid any complications from electrode–skin contact + Easily scalable in terms of sensing channels; this is mainly because of the simplicity of the measuring structure − Higher data processing/transmission requirements; the need for special conductive and/or dielectric materials; relatively complex sensor structures	− Needs to closely adhere to joint moving
I MATERIAL PROPERTIES		
II ENERGY SUPPLY	Electric current	Electric current
III RESOLUTION	40 Hz	
IV SENSITIVITY		Single layer: 6405 Ω for Δθ = 37°
V MEASUREMENT RANGE		Double layer: 5100 Ω for Δθ = 37°
VI TRL	6–8	6–8

Schematic sketch labels:
- Isolation
- Front layer (electrode array)
- ECPC Layer (Carbon polymer, etc.)
- Back layer (electrode array)
- Isolation

	MOBILTECH	
	MOISTURE- AND CHEMICAL-SENSITIVE SENSOR THREAD	**CARBON-FILLED CELLULOSE PHASE**
SENSOR TYPE	Mechanical, chemical	Mechanical
MEASURAND	Visual assessment	Electric current
CONSTRUCTION PRINCIPLE	n/a	Fiber, filament, film
GEOMETRY	Linear	Linear, planar
MATERIAL	n/a	Polymer
Procedural principle	Permanent identification of harmful environmental influences through the use of threads which change their shape, color or volume while absorbing liquids. The core yarn must be UV-resistant and clearly distinguished in color from the load-bearing tape. For the sheath fibers of the yarn, a material must be selected which is changed in shape, color or structure by UV radiation. [4]	Carbon-filled cellulose fiber. Detection of liquids or vapors via electrically conductive filaments from dry-wet spun cellulose dotted with charge carriers (graphite, carbon black, pigments with semiconducting layers, metallic fibers or carbon fibers) whose conductivity changes under tension/pressure or with increasing moisture content. [8]
Schematic sketch	Core yarn (UV sensor) / Coat fiber (wear sensor)	Humidity / Drag / Print / Mantle / C-doped core
Known/possible field of application	A friction-spun sensor thread represents a combination of an abrasion sensor and a UV sensor.	Detection of liquids or vapors.
Possible sensor variants	Decrease in abrasion resistance with increasing exposure to UV radiation.	Mechanically stable even at high temperatures. Sometimes even fire-retardant.
Opportunities and challenges		− Increasing carbon-black content reduces substance strength, ductility and toughness − Doping with carbon black influences the material viscosity to such an extent that stable thread formation is not possible at normal spinning speeds − If the doping with soot is too high, the electrical resistance increases disproportionately
MATERIAL PROPERTIES		
ENERGY SUPPLY	None	Electric current
RESOLUTION		
MATERIAL		
MATERIAL PROPERTIES		
TRL	9	6–8

	MOBILTECH	
	FIBER-OPTIC PH SENSOR	**INTEGRATED OPTICAL FREQUENCY DOUBLER**
1 | SENSOR TYPE	Chemical	Thermal
2 | MEASURAND	Electromagnetic light spectrum	Electromagnetic light spectrum
3 | CONSTRUCTION PRINCIPLE	Fiber	Fiber-optic conductor
4 | GEOMETRY	Linear	Linear
5 | MATERIAL	Polymer, glass	
a) Procedural principle	Utilization of the light absorption dependent on the pH value of the surrounding medium in a fiber-optic probe consisting of a segment of a multimode optical fiber whose end forms the sensor head. In this area, both the coating and the cladding of the fiber are removed, so that a sensitive layer of a copolymer with immobilized dye is polymerized onto the core. Electromagnetic radiation is guided in such a way that the light rays pass through the interface between the fiber core and the sensitive layer and are returned to the core by total reflection at the interface between the sensitive layer and the aqueous analyte. Wavelength-selective absorption occurs. [10]	Determination of absolute temperatures by means of optical frequency doubling, at which a special light wavelength is required for a known temperature of the resonator in order to achieve a frequency conversion (phase matching of fundamental and harmonic wave) with high efficiency. [1]
b) Schematic sketch		
c) Known/possible field of application	Chemical-analytical measurements.	Temperature monitoring of textile structures.
d) Possible sensor variants	Low influence of the internal thickness on the sensor characteristic curve.	Particularly high efficiency.
e) Opportunities and challenges	+ High long-term stability + High sensitivity + Damping arm	− The prerequisite for measurement is a tunable, coherent light source with enough power to operate the resonator
I MATERIAL PROPERTIES	Six months service life	
II ENERGY SUPPLY	Light	Electric current
III RESOLUTION		
IV MATERIAL		
V MATERIAL PROPERTIES	At 680 nm, 0.06 absorbance units per pH unit over the measuring range of four pH units	
VI TRL	9	6–8

	MOBILTECH	
	TEMPERATURE SENSOR	**PYROMETERS**
SENSOR TYPE	Thermal	Thermal
MEASURAND	Electric current, electromagnetic light spectrum, transmitted light, temperature	Electromagnetic light spectrum, temperature
CONSTRUCTION PRINCIPLE	Thread	Fiber-optic conductor
GEOMETRY	Linear	Linear
MATERIAL	Metals, electrically conductive polymers, glass fibers	Sapphire glass, quartz glass
Procedural principle	Design of thread-shaped sensors for the investigation of thermal loads based on low-melting metal wires, which change their electrical properties under thermal load. [4] Temperature determination by measuring the change of the refraction coefficient of the light-guide sheath under temperature change, which leads to a corresponding transmission difference. [9]	Fiber optic measurement method that determines the temperature by analyzing the cavity radiation of a black body. The radiation spectrum of the black body shifts according to Planck's law of radiation depending on temperature. [1]
Schematic sketch	 Heat Low-melting metal	 Electricity Fiber Heat radiation
Known/possible field of application	Temperature monitoring of textile structures.	Non-contact temperature measurement.
Possible sensor variants	Use of threads of electrically conductive polymers or electrically conductive coated polymers. Temperature sensors based on the principle of absorption edge displacement, using filter glasses instead of semiconductor elements.	Very small heat capacity allows measurement of rapid temperature changes.
Opportunities and challenges	+ High reproducibility + Short response time + High accuracy + Low tendency for thread or surface production due to unfavorable properties of the metals	+ Measurement of very high temperatures possible
MATERIAL PROPERTIES		
ENERGY SUPPLY	None	Electric current
RESOLUTION		
MATERIAL		Measurement accuracy of 0.05%
MATERIAL PROPERTIES	50–250 °C	Up to about 2000 °C
TRL	6–8	9

MOBILTECH

	SENSOR THREAD WITH COLOR AND LIGHT EFFECTS	WEAR SENSOR
1 | SENSOR TYPE	Mechanical, chemical	Mechanical
2 | MEASURAND	Visual assessment	Visual assessment
3 | CONSTRUCTION PRINCIPLE		Thread
4 | GEOMETRY	Linear	Linear
5 | MATERIAL		
a) Procedural principle	Permanent signals of loading and wear of the material are visualized without the supply of auxiliary energy by generating the following effects: decomposition of the sensor thread, change in color, shape or volume (swelling, shrinkage, crimping, bending), turbidity or change in mechanical properties (e.g., embrittlement by UV radiation). The preferred design form is the core-sheath structure of friction-spun wrapping yarns, in which after destruction of the sensor material arranged in the sheath a luminous signal thread arranged in the core becomes visible. [4]	Visual assessment of wear by binding colored threads under the fabric surface of tapes and ropes. If wear occurs, the colored threads become visible on the surface. [4]
b) Schematic sketch	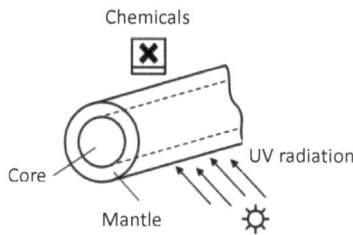 Chemicals Core — Mantle — UV radiation	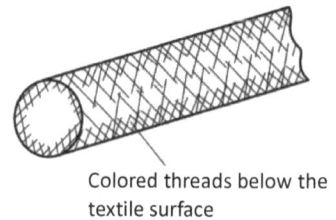 Colored threads below the textile surface
c) Known/possible field of application	Structural health monitoring of ropes.	Structural health monitoring of ropes.
d) Possible sensor variants		
e) Opportunities and challenges		
I MATERIAL PROPERTIES		
II ENERGY SUPPLY		
III RESOLUTION		
IV SENSITIVITY		
V MEASUREMENT RANGE		
VI TRL	6–8	9

	MOBILTECH	
	STRAIN SENSOR	**CONTROL TEAR STRIP**
SENSOR TYPE	Mechanical	Mechanical
MEASURAND	Electric current	Visual assessment, electric current
CONSTRUCTION PRINCIPLE	Thread	Thread
GEOMETRY	Linear, diameter: 0.5–2.5 mm	Linear
MATERIAL	Kevlar, carbon-black-filled silicone rubber	Polyester, silver-plated polyamide, metallic fine wires, cellulose fiber filled with carbon, glass
Procedural principle	Measuring arrangement for determining the strain state in ropes. Based on the location of metal balls incorporated at defined distances by electromagnetic means, the strain results from the distance and the traversing speed of the balls, since these variables are associated with a change in the specific electrical parameters. [4]	Permanent indication of a one-time load overrun of a belt due to the failure of a control tear thread at a defined elongation value which is significantly below the elongation at break of the belt. [4]
Schematic sketch	Distance d / Speed v / Metal sphere / Textiles carrier band / Induction sensor	Tension / Weight breakage
Known/possible field of application	Detection of individual wire breaks in steel ropes, e.g., in kevlar elevator ropes. Use for in situ monitoring and determination of load cycles.	Structural health monitoring of ropes.
Possible sensor variants	Measurement of strains and strain peaks on the basis of a reproducible dependence on strain and electrical resistance, while maintaining the strain state by plastic deformation. [4]	Non-conductive control tear thread: Consists of textile materials such as polyester or polyamide, whose geometric integration into the textile load-handling device is decisive for the elongation of the overall system at which failure occurs. Detections of a few percent can be realized by means of control yarns of non-typical textile elongations such as carbon fiber, glass fiber or Twaron aramid filament yarn.
Opportunities and challenges	+ For protection against overloading, it is not necessary for the sensor thread to fail. Exceeding a defined strain state is sufficient for the output of an alarm signal + By also detecting strain peaks, strain sensors open up a wide range of applications, from crack sensors to sensors for detecting strain peaks − Process cannot be applied to man-made fiber tapes and ropes	+ Silver-coated polymer thread: unsuitable as electrically conductive control tearing thread, since elongations at break cannot be reproduced or the parallel position of the untwisted filaments results in only individual filaments tearing in case of failure and the applied tension remaining constant − Metallic fine wire: very sensitive to breakage, otherwise excessively high elongation at break compared to load-bearing agent − Cellulose fiber with carbon filling: lower, moisture-dependent conductivity than silver-plated polyamide yarns or fine wires. − Optically conductive control thread: buckling sensitivity, critical mechanical behavior.
MATERIAL PROPERTIES	Hardness: 50 ± 5 Shore A; density: 1.13 g/cm^3; tear strength: ≥3.5 N/mm^2; elongation at break: ≥200%; specific volume resistance: ≤12 Ωcm	
ENERGY SUPPLY	Electric current	Electric current
RESOLUTION		
MATERIAL		
MATERIAL PROPERTIES		
TRL	9	9

	MOBILTECH	
	FRICTION-SPUN ABRASION SENSOR THREAD	**ADAPTIVE FIBER COMPOSITES (ADAPTRONICS)**
1 | SENSOR TYPE	Mechanical	Mechanical
2 | MEASURAND	Visual assessment	Vibration, deformation
3 | CONSTRUCTION PRINCIPLE	Thread	
4 | GEOMETRY	Linear	
5 | MATERIAL	Polypropylene, polyethylene terephthalate	
a) Procedural principle	Sensor with optical signal output in the event of critical wear or damage to the outer sheath of a load-bearing rope or tape. The sensor thread has a core-sheath structure, the signal-colored core being sheathed with thermoplastic staple fiber. This has the color of the load-bearing textile and is integrated into its outer shell in such a way that it is exposed to abrasion during use. [4]	Active vibration suppression by piezoelectric films and fibers which self-adjust to changing component vibrations and deformations by integrated sensors as well as initiate counter-signals via actuators into the textile structure. [20]
b) Schematic sketch		

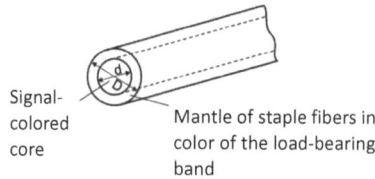

Signal-colored core

Mantle of staple fibers in color of the load-bearing band

c) Known/possible field of application	Structural health monitoring of ropes.	
d) Possible sensor variants	Variation of the ratio of core to shell diameter. Core yarn made of PET, sheath yarn made of PP; core thread not signal-colored, but made of fluorescent material for UV detection. Variation of core and sheath strength. Core yarn made of PP, sheath yarn made of PET.	Lightweight construction possible; high-stiffness and high-strength fiber composites.
e) Opportunities and challenges	+ The use of a fluorescing signal thread in the thread core enables an automated visual inspection of the wear condition by means of camera technology, even for soiled or very colorful load-bearing textiles + With increasing sheath fineness, there is a significant increase in bearable double chafing	− Fundamentally low mechanical resistance to noise and vibration
I MATERIAL PROPERTIES		
II ENERGY SUPPLY	None	None
III RESOLUTION		
IV SENSITIVITY		
V MEASUREMENT RANGE		
VI TRL	6–8	9

	MOBILTECH	
	STRAIN/PRESSURE SENSOR	**INTELLIGENT MEMBRANE**
SENSOR TYPE	Mechanical	Mechanical
MEASURAND	Electric current	Electromagnetic light spectrum, electric current, noise level
CONSTRUCTION PRINCIPLE	Weft knit, warp knit	Fiber bundle in woven or knitted structure
GEOMETRY	Planar	Areal, change of shape up to 8 times its size
MATERIAL	Stainless steel	Nickel titanium alloy
Procedural principle	Spacer weft-knit made of electrically conductive stainless-steel-fiber yarns for detecting the position of the contact and the size of the contacting surface when the specific electrical resistance of the electrically conductive conductor paths changes as a result of elongation or pressure. [21]	A membrane with built-in sensors which reacts to stimuli such as light, contact, noise or environmental movements in a mobile manner via muscle wires made of Ni-Ti alloy developing different temperatures at certain currents and passing through different movements. [22]
Schematic sketch	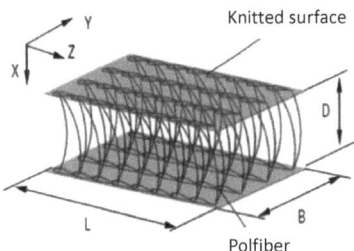	
Known/possible field of application	Flat pressure load in buildings.	
Possible sensor variants	Spacer warp-knit can also be used instead of spacer weft-knit.	Use in any size possible.
Opportunities and challenges	– Spacer warp-knit has a hysteretic force behavior and is therefore less suitable as a pressure sensor	– Very expensive materials
MATERIAL PROPERTIES		
ENERGY SUPPLY	Electric current	Electric current
RESOLUTION		
MATERIAL		
MATERIAL PROPERTIES		
TRL	6–8	<6

MOBILTECH

	PRESSURE SENSOR	THREE-DIMENSIONAL SPACER WARP-KNIT
1 | SENSOR TYPE	Mechanical	Mechanical
2 | MEASURAND	Electric current, pressure	Electric current
3 | CONSTRUCTION PRINCIPLE	Elastic weft-knit or warp-knit equipped with electrically conductive threads	Warp knit
4 | GEOMETRY	Dimensions of the pressure sensor: 10 mm x 10 mm x 1 mm	Planar
5 | MATERIAL	Electrically conductive metals	
a) Procedural principle	Pressure sensor with strip-like or filament-like elements which each have a layered structure and are electrically conductive. When pressure is applied, the layers touch each other and a closed circuit is formed which indicates the pressure. [47]	Three-dimensional spacer warp-knit with integrated ultrasonic sensors for monitoring body movement. [50]
b) Schematic sketch		
c) Known/possible field of application	Clothing for monitoring heart activity and recording skin resistance, perspiration and body temperature.	
d) Possible sensor variants	Pressure sensitive stocking. [49]	Flexible elastane material guarantees flexibility and wearing comfort.
e) Opportunities and challenges		
I MATERIAL PROPERTIES	Environmentally stable at 0–50 °C, 30–90% relative humidity	
II ENERGY SUPPLY	Electric current	Electric current
III RESOLUTION		
IV SENSITIVITY		
V MEASUREMENT RANGE	Surface pressure: 0–10 kg/cm^2	
VI TRL	6–8	6–8

Schematic sketch labels: Electrically conductive fiber · Poorly conductive polymer shell · Polymer fiber · **Longitudinal section**

	MOBILTECH	
	INTELLIGENT SKIN ARCHITECTURE	**GYROSCOPE (ROTATION SENSOR)**
SENSOR TYPE	Mechanical	Mechanical
MEASURAND		Electric current
CONSTRUCTION PRINCIPLE	Weave	Fiber optic
GEOMETRY	Planar	Linear conductor, fiber length between 100 and 1000 m
MATERIAL		
Procedural principle	Optical fibers woven into a carrier material which serve as sensors for optical information transmission. [53]	Ring interferometer which evaluates the phase difference between the opposing light waves, which is dependent on the angular velocity, as a measured variable. Polarized laser light passes between two beam splitters before it is coupled into the two ends of the same fiber coil. In the case of a stationary system, light paths of equal lengths of the circulating modes result in a constructive interference at the output of the second beam splitter, whereas a destructive interference occurs at the output of the first beam splitter. The relativistic Sagnac effect results in a phase difference $\Delta\Phi$ between the light waves rotating in opposite directions, which is proportional to the product of the conversion number m and the enclosed area A. [1]
Schematic sketch		
Known/possible field of application	Acquisition of data; image processing; communication.	Earth rotation measurement. Navigation tools. Robot control.
Possible sensor variants	Supporting weaving of the optical fibers into channels. Arrangement of the optical fibers in a grid-like mat consisting of fibers of any carrier material. Woven structure comprising a first group of warp-direction yarns and a second group of weft-direction yarns with optical fibers arranged between selected pairs of the first group. Optoelectronic packaging structure with two sections, in each of which the abovementioned woven structure is placed.	Integrated optical resonator: sensitivities up to several 100s of °/h.
Opportunities and challenges	+ Low construction volume; low weight + High tensile strength; high elasticity; high resistance to weathering; high resistance to chemicals; high tear strength; high dimensional stability; high wear resistance − Sensitivity to deflection, leading to a deterioration in transmittance	+ Miniaturization of the fiber-optic gyroscope through integrated optics + Use in areas with short-term stability as well as with required long-term stability possible
MATERIAL PROPERTIES		
ENERGY SUPPLY	None	Laser light
RESOLUTION		
SENSITIVITY		Up to 3–10 °/h
MEASUREMENT RANGE		
TRL	6–8	9

MOBILTECH

	SENSOR HEATING ELEMENT	OVERSTITCHED SYSTEM
1 | SENSOR TYPE	Mechanical, thermal	Mechanical
2 | MEASURAND	Electric current	Electric current
3 | CONSTRUCTION PRINCIPLE	Fleece	Spacer warp-knit consisting of two metallized fabrics and spacer material (warp knit or fleece)
4 | GEOMETRY	Planar	Linear, planar
5 | MATERIAL	Carrier film: polymide (PI), PET, PEN; conductor tracks: copper; protective layer: plastic or fleece	Electrically conductive metals
a) Procedural principle	Flexible sensor unit for detecting seat occupancy in a passenger car, which is in the form of a conductive film laminated onto a carrier material. The heating conductors are arranged between the conductor tracks of the film. A change in the pressure on or temperature of the sensor tracks is accompanied by a change in their electrical properties (electrical resistance), which in turn influences the electric current to be measured. [61]	System having at least two flat conductors formed from metallized fabric, which are electrically insulated from one another by a spacer material (warp-knit or fleece) and conduct electricity only o contact. The spacer material is used for electrical insulation of the electrical conductors, and for the seat and climatic comfort of the driver. [62]
b) Schematic sketch		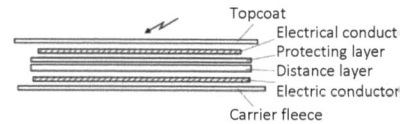
c) Known/possible field of application		Capacitive occupancy detection system integrated the seat of a motor vehicle.
d) Possible sensor variants	Realization by pressure or temperature sensors. Design of the seat occupancy detection sensor for deactivation/activation of airbags in the automotive sector. Design of the seat occupancy detection sensor to detect a pressure profile that is correlated with the heat output.	
e) Opportunities and challenges	+ Simplified sensor design + Simplified manufacturing process by reducing the number of process steps in a combination of sensor and heating element + Reduced number of components + Cost saving	+ Simple manufacturing process of a system of an given size in only one process + Minimized risk of an electrical short-circuit durin the sewing process due to elastic overstitch protection layer
I MATERIAL PROPERTIES		
II ENERGY SUPPLY	Electric current	None
III RESOLUTION		
IV SENSITIVITY		
V MEASUREMENT RANGE		
VI TRL	9	9

Schematic sketch labels: Topcoat, Electrical conduct, Protecting layer, Distance layer, Electric conductor, Carrier fleece

	MOBILTECH	
	PRESSURE SENSOR	**FIBER-COATED SENSORS**
SENSOR TYPE	Chemical	Chemical
MEASURAND	Electromagnetic light spectrum	Electromagnetic light spectrum
CONSTRUCTION PRINCIPLE		Weave
GEOMETRY	Linear	Planar
MATERIAL		FBG fibers in fabric
Procedural principle	Realization of a pressure sensor with the help of two aligned optical waveguides (one fixed, the other movable).	FBG is a distributed Bragg reflector constructed in a short segment of optical fiber that reflects certain wavelengths of light and transmits all others. This is achieved by creating a periodic variation of the refractive index of the fiber core, which generates a wavelength-specific dielectric mirror. [23]
Schematic sketch	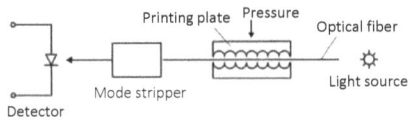	
Known/possible field of application		Used in seismology, pressure sensors for extremely harsh environments, and downhole sensors in oil and gas wells for measurement of the effects of external pressure, temperature, seismic vibrations and inline flow measurement.
Possible sensor variants	Pressure measurement using the "microbending-effect", in which small deviations of the optical fiber axis from a straight line cause mechanical stresses in the core and cladding, which in turn cause light to be decoupled.	Integration of Bragg fiber as warp thread; into a 3D woven; embedded in a conveyor belt; inserted into a groove and threaded into flat-woven fabric.
Opportunities and challenges		+ Inline optical filter to block certain wavelengths, or as a wavelength-specific reflector
MATERIAL PROPERTIES		
ENERGY SUPPLY	Light	Light
RESOLUTION		
SENSITIVITY	0–20 bar	
MEASUREMENT RANGE		
TRL	6–8	<6

	HUMIDITY SENSOR	OPTOELECTRONIC SENSOR
1 | SENSOR TYPE	Chemical	Chemical
2 | MEASURAND	Electric current	Electromagnetic light spectrum
3 | CONSTRUCTION PRINCIPLE	Weft knit	Fiber
4 | GEOMETRY	Planar	Linear
5 | MATERIAL	Electrically conductive yarn	Plexiglas
a) Procedural principle	Knitted fabric with a basic weft-knit which contains at least one thread made of a material which changes its electrical resistance when affected by moisture. The weft knit is equipped with an integrated moisture sensor consisting of at least two electrodes arranged at a distance, which are electrically connected to each other in case of moisture. [2]	Detection of adhering liquid components in or on liquid-storing substances by detecting the change in the transmission of light in a light-guide with the liquid component to be taken. [3]
b) Schematic sketch	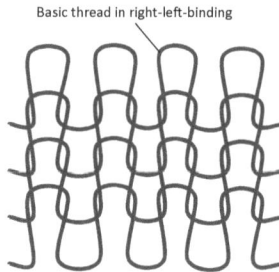 Basic thread in right-left-binding	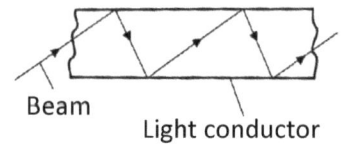 Beam — Light conductor
c) Known/possible field of application	Woven fabrics in which electrically well conducting and electrically not well conducting threads are alternately woven with each other. Electrical connection means in the form of terminals, plug-in connection parts.	Detection of liquid content of soils, textiles or granulates. Monitoring tasks, for example in landfills.
d) Possible sensor variants	Electrical means of connection can be connected to the monitoring station via textile conductors. The textile behavior ensures that the joint is extremely flexible and elastic.	Cost-effective.
e) Opportunities and challenges	+ Integration of the sensor directly into the garment, with no external application necessary	
I MATERIAL PROPERTIES		
II ENERGY SUPPLY	Electric current	Light
III RESOLUTION		
IV MATERIAL		
V MATERIAL PROPERTIES		
VI TRL	6–8	9

	ECOTECH	
	MOISTURE- AND CHEMICAL-SENSITIVE SENSOR THREAD	**WATER DETECTOR**
SENSOR TYPE	Mechanical, chemical	Chemical
MEASURAND	Visual assessment	Visual assessment
CONSTRUCTION PRINCIPLE	n/a	Textile tape, thread, thread bundle, textile fiber composite, fleece, paper, film, wire, warp knit
GEOMETRY	Linear	Punctiform, linear, planar, voluminous
MATERIAL	n/a	Cellulose, polyolefin, nylon, Nomex, Teflon, plastic, polyester, ceramic, metal, wool
Procedural principle	Permanent identification of harmful environmental influences through the use of threads which change their shape, color or volume while absorbing liquids. The core yarn must be UV-resistant and clearly distinguished in color from the load-bearing tape. For the sheath fibers of the yarn, a material must be selected which is changed in shape, color or structure by UV radiation. [4]	Textile probe with sufficiently large stored active substance depot, which on contact with the substance to be investigated causes a visual chemical change in the detector depending on the composition and movement of the analyte. The change occurs in the form of a substance solution, substance deposition or formation of a new substance at the detector itself. [5]
Schematic sketch		
Known/possible field of application	A friction-spun sensor thread represents a combination of an abrasion sensor and a UV sensor.	Analysis of gas and water, and also soil and sediment, samples.
Possible sensor variants	Decrease in abrasion resistance with increasing exposure to UV radiation.	The resistance of the optically visually-recognizable color pattern of the detector to water with a different composition to that of the measuring point and the atmosphere, which is exposed to short-term effects, prevents falsification of the measurement.
Opportunities and challenges		
MATERIAL PROPERTIES		
ENERGY SUPPLY	None	None
RESOLUTION		>1 h
MATERIAL		
MATERIAL PROPERTIES		
TRL	9	9

Core yarn
(UV sensor)

Coat fiber
(wear sensor)

	DETECTION MEANS	FIBER-OPTIC SENSOR
1 | SENSOR TYPE	Chemical	Chemical
2 | MEASURAND	Visual assessment	Electromagnetic light spectrum
3 | CONSTRUCTION PRINCIPLE	Fiber	Fiber
4 | GEOMETRY	Linear, planar	Linear
5 | MATERIAL	Cellulose, plastic, glass, ceramics	Cotton for protective vision, fluoride glass for light-guide sheath and core
a) Procedural principle	Detection of substances with shaped and unshaped detection means, containing fibers and/or adhesives which react to environmental influences via a color change, which serves as an indicator. [6]	Fiber-optic sensor for detecting gaseous or liquid media, surrounded by an optical fiber sheath consisting of a fluoride glass of low chemical resistance to be detected on contact with the analyte, decomposition of the sheath takes place within a characteristic chemically induced reaction time until the sensor responds as a function of the original thickness of the sheath, the temperature and the concentration of the attacking medium while maintaining the total reflection condition (lower refractive index of the sheath with respect to the optical fiber core). A hygroscopic textile protective layer around the light-guide sheath increases the corrosive effect of the attacking medium on the light-guide sheath. [7]
b) Schematic sketch		Fiber-optic core / Fiber-optic sheath / Gas- and liquid-permeable protective cover
c) Known/possible field of application	Analysis of water, soil and sediment samples of natural and artificial constituents including radioactive contaminants. Control measures in food and feed production. Production and monitoring of industrial products, including gases. Monitoring and control of industrial processes. Control measures in the nuclear sector.	Detection of gaseous and liquid media. Monitoring of electrical cables and lines, as well as endangered installations, pipelines, equipment and buildings for the ingress of water, water vapor, acid, alkalis or other gases and liquids.
d) Possible sensor variants	Spatially and temporally seamless qualitative monitoring and documentation of processes possible.	High mechanical strength.
e) Opportunities and challenges	− The detection medium can also be used to a limited extent as a filter for certain substances	+ High response sensitivity, even to individual media only + Targeted analysis of individual specific substances with desired concentration content + Low manufacturing and general cost
I MATERIAL PROPERTIES		Light-guide sheath with lower refractive index than conductor core, light-guide sheath made of fluoride glass with lower hydrolytic resistance
II ENERGY SUPPLY	None	None
III RESOLUTION		
IV MATERIAL		
V MATERIAL PROPERTIES		
VI TRL	9	6–8

	ECOTECH	
	PH SENSOR	**FIBER OPTIC SENSOR**
SENSOR TYPE	Chemical	Chemical
MEASURAND	Visual assessment	Electromagnetic light spectrum
CONSTRUCTION PRINCIPLE		Fiber
GEOMETRY	Linear	Linear
MATERIAL		Cotton for protective vision, fluoride glass for light-guide sheath and core
Procedural principle	Measurement of substance concentrations, which are not directly accessible spectroscopically, with a sensitive chemoreceptor. This receptor is a sensor, at the end of which a specific indicator (e.g., phenol red in polyacrylamide) is immobilized, by which a change in pH is measured either in reflection or as fluorescence. [9]	Fiber-optic sensor for detecting gaseous or liquid media, surrounded by an optical fiber sheath consisting of a fluoride glass of low chemical resistance to be detected on contact with the analyte, decomposition of the sheath takes place within a characteristic chemically induced reaction time until the sensor responds as a function of the original thickness of the sheath, the temperature and the concentration of the attacking medium while maintaining the total reflection condition (lower refractive index of the sheath with respect to the optical fiber core). A hygroscopic textile protective layer around the light-guide sheath increases the corrosive effect of the attacking medium on the light-guide sheath. [7]
Schematic sketch	Optical fiber / Optical fiber / Immobilized indicator / Permeable membrane	Fiber-optic core / Fiber-optic sheath / Gas- and liquid-permeable protective cover
Known/possible field of application		Detection of gaseous and liquid media. Monitoring of electrical cables and lines, as well as endangered installations, pipelines, equipment and buildings for the ingress of water, water vapor, acids, alkalis or other gases and liquids.
Possible sensor variants	Very accurate pH measurement only achievable for very small ranges (approximately three pH units).	High mechanical strength.
Opportunities and challenges		+ High response sensitivity, even to individual media only + Targeted analysis of individual specific substances with desired concentration content + Low manufacturing and general cost
MATERIAL PROPERTIES		Light-guide sheath with lower refractive index than conductor core, light-guide sheath made of fluoride glass with lower hydrolytic resistance
ENERGY SUPPLY	None	None
RESOLUTION		
MATERIAL		
MATERIAL PROPERTIES	0.005 pH units	
TRL	6–8	6–8

PAGE 168	INTEGRATED OPTICAL FREQUENCY DOUBLER	INTEGRATED OPTICAL RESONATOR
1 | SENSOR TYPE	Thermal	Thermal
2 | MEASURAND	Electromagnetic light spectrum	Electromagnetic light spectrum
3 | CONSTRUCTION PRINCIPLE	Fiber-optic conductor	
4 | GEOMETRY	Linear	Linear
5 | MATERIAL		$LiNbO_3$
a) Procedural principle	Determination of absolute temperatures by means of optical frequency doubling, in which a special light wavelength is required for a known temperature of the resonator in order to achieve a frequency conversion (phase matching of fundamental and harmonic wave) with high efficiency. [1]	The temperature changed by means of optical resonators integrated in LiNbO3 with a periodic characteristic curve. To be able to record the number of orders passed as a function of the direction of the phase (or temperature) change requires two signals phase-shifted by 90°. It is advantageous to use the output signals to arrive at an evaluation, which counts in each case with the zero crossing, and thus an independence from slow fluctuations of the light intensity is obtained. The phase modulation required for differentiation is achieved by frequency modulation of the laser light or by electro-optical modulation of the optical path length of the resonator. [1]
b) Schematic sketch	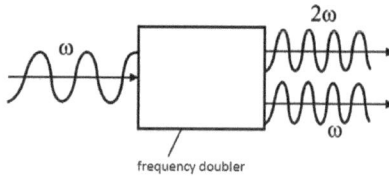 frequency doubler	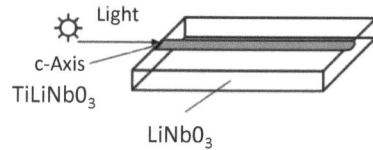
c) Known/possible field of application	Temperature monitoring of textile structures.	Temperature monitoring of textile structures.
d) Possible sensor variants	Particularly high efficiency.	The sensitivity of the temperature sensor can be determined in wide ranges by the length of the textile component and the wavelength of the light.
e) Opportunities and challenges	− The prerequisite for measurement is a tunable coherent light source with enough power to operate the resonator	+ Simple measuring system with high accuracy when supplying the resonator sensor element via a polarization-maintaining monomode fiber + Measurement of smallest temperature changes possible due to the strong temperature dependence of the refractive index − Measurement of absolute temperatures not possible
I MATERIAL PROPERTIES		
II ENERGY SUPPLY	Electric current	None
III RESOLUTION		
IV MATERIAL		Sensitivity of 35 impulses/K, resolution of 29 impulses/K
V MATERIAL PROPERTIES		
VI TRL	6–8	6–8

	ECOTECH	
	TEMPERATURE SENSOR	**FIBER-OPTIC TEMPERATURE SENSOR**
SENSOR TYPE	Thermal	Thermal
MEASURAND	Electric current, electromagnetic light spectrum, transmitted light, temperature	Electromagnetic light spectrum, temperature
CONSTRUCTION PRINCIPLE	Thread	Fiber processed into fabric
GEOMETRY	Linear	Flat, length of one optical waveguide up to 20 m
MATERIAL	Metals, electrically conductive polymers, glass fibers	Fiber made of glass, sheet material, e.g., geotextile
Procedural principle	Design of thread-shaped sensors for the investigation of thermal loads based on low-melting metal wires, which change their electrical properties under thermal load. [4] Temperature determination by measuring the change of the refraction coefficient of the light-guide sheath under temperature change, which leads to a corresponding transmission difference. [9]	Textile temperature-measuring mat with meandering optical waveguide for checking and monitoring the insulation of cladding pipe sections. The temperature is measured via a fiber-optic recording of the temperature-dependent anti-Stokes line in the optical waveguide. The temperature can be measured either continuously or sequentially by evaluating the scattered light pulses depending on the run time. From the registered temperature curve, the effectiveness of the insulation can be concluded. [11]
Schematic sketch	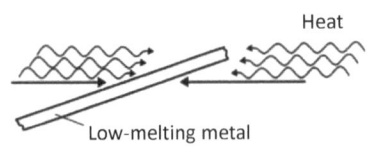 Heat Low-melting metal	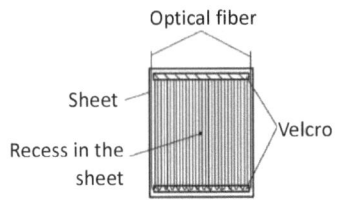 Optical fiber Sheet Recess in the sheet Velcro
Known/possible field of application	Temperature monitoring of textile structures.	Control and monitoring of the insulation of pipe sections.
Possible sensor variants	Use of threads of electrically conductive polymers or electrically conductive coated polymers. Temperature sensors based on the principle of absorption edge displacement using filter glasses instead of semiconductor elements.	Replacement of the fiber-optic cable by flat distributed single sensors.
Opportunities and challenges	+ High reproducibility + Short response time + High accuracy + Low tendency for thread or surface production due to unfavorable	+ Simultaneous temperature measurement of several locations by means of a light pulse and the dependence of the temperature on the propagation time of the light + Temperature measurement already possible during the manufacture of the pipe insulation + Cost-effective method, since one fiber-optic cable is sufficient for temperature measurement in principle + Location-dependent measurement enables local weak points in the pipe insulation to be detected
MATERIAL PROPERTIES		Fabrics not subject to tensile, compressive and tear loads
ENERGY SUPPLY	None	Electric current
RESOLUTION		
MATERIAL		0.1 K
MATERIAL PROPERTIES	50–250 °C	100–750 K
TRL	6–8	9

	ECOTECH	
	FIBER-OPTIC DISPLACEMENT TRANSDUCER	**ACTIVE FIBER-OPTIC SENSOR**
1 | SENSOR TYPE	Mechanical	Chemical
2 | MEASURAND	Path, route	Visual assessment
3 | CONSTRUCTION PRINCIPLE	Fiber-optic conductor	Fiber
4 | GEOMETRY	Linear	Linear
5 | MATERIAL		
a) Procedural principle	Measurement of paths on the basis of various principles. In particular, fiber optic measurements of a large number of physical quantities that can be converted into paths by test specimens. [1]	Measurement of the distance between sensor and fluid environment, the concentration of chemicals in the fluid environment, the pH value of aqueous solutions, and the partial pressures of a gas by evaluating the light transmitted via the fiber-optic laser, if this changes characteristically as a reaction between sensor reagent and surrounding environment. [12]
b) Schematic sketch		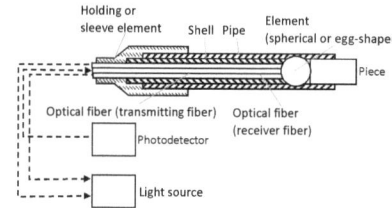
c) Known/possible field of application	Measurement technology, from displacement measurement, angle, pressure or acceleration can also be measured, depending on the arrangement.	Control of chemical processes in nuclear and industrial areas, underground nuclear waste in the environment, in medical and biological analysis, as well as in the agri-food industry; medical applications; biochemical applications; use in the food industry.
d) Possible sensor variants		Fiber optically active sensor. [13]
e) Opportunities and challenges		+ Long service life + Simple sterilization + High stability − Limited pH measuring range − Limited reproducibility of the reaction between optical fibers and the immobilized reagent
I MATERIAL PROPERTIES		Bulky sensor material
II ENERGY SUPPLY		Light
III RESOLUTION		
IV SENSITIVITY		
V MEASUREMENT RANGE	10^{-10}–1 m	
VI TRL	9	9

Schematic sketch labels: Holding or sleeve element, Shell, Pipe, Element (spherical or egg-shape), Piece, Optical fiber (transmitting fiber), Optical fiber (receiver fiber), Photodetector, Light source

	ECOTECH	
	SOUND SENSOR (HYDROPHONE)	**RAPID-SHRINK FIBER**
SENSOR TYPE	Mechanical	Chemical
MEASURAND	Electric current	Visual assessment
CONSTRUCTION PRINCIPLE	Fiber-optic conductor	Warp knit, weave
GEOMETRY	Linear	Linear, planar
MATERIAL	Quartz glass	Elastomer (polyutherane, rubber)
Procedural principle	Fiber-optic hydrophone (Mach–Zehnder interferometer) for highly sensitive detection of pressure differences between measuring and reference fibers. By modulating the refractive index of the measuring fiber, the sound pressure changes the phase length of the passing light and thus the interference signal, which is detected by two photodiodes and fed to the amplifier via a high-pass filter. The signal behind the low pass is used to stabilize the operating point of the interferometer against slow fluctuations, e.g., due to temperature changes. [1]	A polymer fiber which shrinks rapidly at ordinary temperature and in contact with water, but retains the fiber shape (impact strength), has high absorbency and has performance characteristics such as rubber elasticity. [14]
Schematic sketch	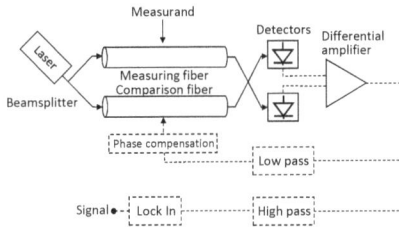	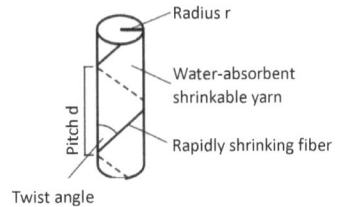
Known/possible field of application	Metrology.	Disposable diapers; fastening tapes; cloths as covers for dampening units in offset printers; cords or cylinders for plant cultivation; cords and nets for the food industry; bank reinforcements.
Possible sensor variants	Due to the flexibility of the quartz glass fibers, sensors with directional characteristics can be manufactured.	A water-absorbing, shrinkable yarn produced by blending or blending spinning the rapidly shrinking fiber and a fiber that shrinks slower than said fiber upon absorption of water. A water-absorbing shrinkable material which consists of a water-absorbing shrinkable fibrous web and a water-absorbent shrinkable yarn that absorbs water at a higher rate and to a greater extent than the fibrous web, with the water-absorbent shrinkable yarn containing the rapidly shrinking fiber.
Opportunities and challenges		
MATERIAL PROPERTIES		At 20 °C, maximum percentage shrinkage >30%
ENERGY SUPPLY	Laser light	
RESOLUTION		0–10 s
MATERIAL		
MATERIAL PROPERTIES		At 20 °C, shrinkage stress = 0.351–1.755 kg/m² (30–150 mg/den)
TRL	9	9

ECOTECH

	PYROMETERS	UV SENSOR FIBER
1 | SENSOR TYPE	Thermal	Chemical
2 | MEASURAND	Electromagnetic light spectrum, temperature	Visual assessment
3 | CONSTRUCTION PRINCIPLE	Fiber-optic conductor	Thread
4 | GEOMETRY	Linear	Linear
5 | MATERIAL	Sapphire glass, quartz glass	
a) Procedural principle	Fiber-optic measurement method that determines the temperature by analyzing the cavity radiation of a black body. The radiation spectrum of the black body shifts according to Planck's law of radiation depending on temperature. [1]	Permanent signaling of the reaching or exceeding of a maximum permissible limit for the effect of UV radiation on load-bearing belts and ropes by accumulation sensors. In contrast to photochromic materials, which only record the instantaneous radiation intensity, accumulation sensors visualize the total measure of the radiation effect. [4]
b) Schematic sketch	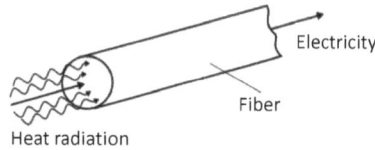 Electricity / Fiber / Heat radiation	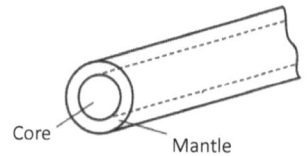 Core / Mantle
c) Known/possible field of application	Non-contact temperature measurement.	
d) Possible sensor variants	Very small heat capacity allows the measurement of rapid temperature changes.	Core-sheath structures in the form of friction and wrapping yarns, which consist of a UV-sensitive sheath (sensor thread) and a luminous signal thread in the core analogous to the abrasion-sensitive sensor threads. Twisted yarns consisting of two or more threads with almost identical (colorimetrically adjusted) hues but different light fastness, which change their appearance from self-colored to multicolored after UV irradiation by bleaching of the threads with lower light fastness.
e) Opportunities and challenges	+ Measurement of very high temperatures possible	+ Semi-quantitative determination of the radiation dose using the reference filament + The elimination of twine production in one additional operation means that the titre of the individual yarns can also be adjusted to the yarn used in the product + Both sensor thread and reference thread can be processed individually in adjacent positions in the tape fabric or braid, provided they are suitable for the weave
I MATERIAL PROPERTIES		
II ENERGY SUPPLY	Electric current	Electromagnetic radiation
III RESOLUTION		
IV SENSITIVITY	Measurement accuracy of 0.05%	
V MEASUREMENT RANGE	Up to about 2000 °C	
VI TRL	9	9

	ECOTECH	
	STRAIN/PRESSURE SENSOR	**CLOTHING INDICATOR FOR UV RADIATION AND OZONE**
SENSOR TYPE	Mechanical	Chemical
MEASURAND	Electric current	Visual assessment
CONSTRUCTION PRINCIPLE	Weft knit, warp knit	
GEOMETRY	Planar	
MATERIAL	Stainless steel	
Procedural principle	Spacer weft-knit made of electrically conductive stainless-steel-fiber yarns for detecting the position of the contact and the size of the contacting surface when the specific electrical resistance of the electrically conductive conductor paths changes as a result of elongation or pressure. [21]	Small-scale application of a variety of fashionable design forms whose color change is accompanied by influencing factors from the environment and at least semi-quantitatively correlates with the hazard potential. The measurement is carried out either by the iodine method, acetone decomposition, oxalic acid decomposition or an IG dosimeter.
Schematic sketch	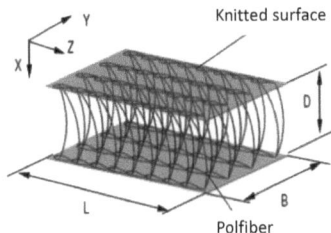	
Known/possible field of application	Flat pressure load in buildings.	Sensor application to swimwear, leisurewear and workwear for outdoor activities for detection of UV radiation.
Possible sensor variants	Spacer warp-knit can also be used instead of spacer weft-knit.	Determination of the intensity of UV radiation by measurement using iodine method, acetone decay, oxalic acid decomposition or IG dosimeter.
Opportunities and challenges	− Spacer warp-knit has a hysteretic force behavior and is therefore less suitable as a pressure sensor	+ Good resistance of the textile carrier to the hazard potential, sensitization technology and reactions causing color change − Doubts as to whether the concentration and intensity of the hazard potential is sufficient to initiate the chemical reaction on the textile
MATERIAL PROPERTIES		
ENERGY SUPPLY	Electric current	Light
RESOLUTION		
SENSITIVITY		
MEASUREMENT RANGE		
TRL	6–8	<6

In the schematic sketch: Y, Z, X axes; Knitted surface; D; L; B; Polfiber

	PRESSURE SENSOR	GYROSCOPE (ROTATION SENSOR)
1 | SENSOR TYPE	Mechanical	Mechanical
2 | MEASURAND	Electric current, pressure	Electric current
3 | CONSTRUCTION PRINCIPLE	Elastic weft-knit or warp-knit equipped with electrically conductive threads	Fiber optic
4 | GEOMETRY	Dimensions of the pressure sensor: 10 mm x 10 mm x 1 mm	Linear conductor, fiber length between 100 and 1000 m
5 | MATERIAL	Electrically conductive metals	
a) Procedural principle	Pressure sensor with strip-like or filament-like elements which each have a layered structure and are electrically conductive. When pressure is applied, the layers touch each other and a closed circuit is formed which indicates the pressure. [47]	Ring interferometer which evaluates the phase difference between the opposing light waves, which is dependent on the angular velocity, as a measured variable. Polarized laser light passes between two beam splitters before it is coupled into the two ends of the same fiber coil. In the case of a stationary system, light paths of equal lengths of the circulating modes result in a constructive interference at the output of the second beam splitter, whereas a destructive interference occurs at the output of the first beam splitter. The relativistic Sagnac effect results in a phase difference $\Delta\Phi$ between the light waves rotating in opposite directions, which is proportional to the product of the conversion number m and the enclosed area A. [1]
b) Schematic sketch		

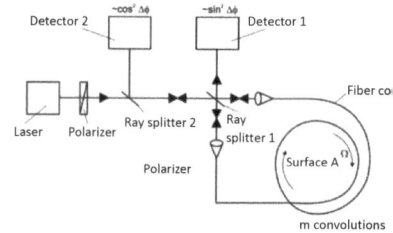

	PRESSURE SENSOR	GYROSCOPE (ROTATION SENSOR)
c) Known/possible field of application	Clothing for monitoring heart activity and recording skin resistance, perspiration and body temperature.	Earth rotation measurement. Navigation tools. Robot control.
d) Possible sensor variants	Pressure sensitive stocking. [49]	Integrated optical resonator: sensitivities up to several 100s of °/h.
e) Opportunities and challenges		+ Miniaturization of the fiber-optic gyroscope through integrated optics + Use in areas with short-term stability as well as with required long-term stability possible
I MATERIAL PROPERTIES	Environmentally stable at 0–50 °C, 30–90% relative humidity	
II ENERGY SUPPLY	Electric current	Laser light
III RESOLUTION		
IV SENSITIVITY		Up to 3–10 °/h
V MEASUREMENT RANGE	Surface pressure: 0–10 kg/cm^2	
VI TRL	6–8	9

	PACKTECH	
	RAPID-SHRINK FIBER	**CONTROL TEAR STRIP**
SENSOR TYPE	Chemical	Mechanical
MEASURAND	Visual assessment	Visual assessment, electric current
CONSTRUCTION PRINCIPLE	Warp knit, weave	Thread
GEOMETRY	Linear, planar	Linear
MATERIAL	Elastomer (polyutherane, rubber)	Polyester, silver-plated polyamide, metallic fine wires, cellulose fiber filled with carbon, glass
Procedural principle	A polymer fiber which shrinks rapidly at ordinary temperature and in contact with water, but retains the fiber shape (impact strength), has high absorbency and has performance characteristics such as rubber elasticity. [14]	Permanent indication of a one-time load overrun of a belt due to the failure of a control tear thread at a defined elongation value which is significantly below the elongation at break of the belt. [4]
Schematic sketch	Radius r, Water-absorbent shrinkable yarn, Rapidly shrinking fiber, Pitch d, Twist angle	Tension, Weight breakage
Known/possible field of application	Disposable diapers; fastening tapes; cloths as covers for dampening units in offset printers; cords or cylinders for plant cultivation; cords and nets for the food industry; bank reinforcements.	Structural health monitoring of ropes.
Possible sensor variants	A water-absorbing, shrinkable yarn produced by blending or blending spinning the rapidly shrinking fiber and a fiber that shrinks slower than said fiber upon absorption of water. A water-absorbing shrinkable material which consists of a water-absorbing shrinkable fibrous web and a water-absorbing shrinkable yarn that absorbs water at a higher rate and to a greater extent than the fibrous web, with the water-absorbent shrinkable yarn containing the rapidly shrinking fiber.	Non-conductive control tear thread: Consists of textile materials such as polyester or polyamide, whose geometric integration into the textile load-handling attachment is decisive for the elongation of the overall system at which failure occurs. Detections of a few percent can be realized by means of control yarns of non-typical textile elongations such as carbon fiber, glass fiber or Twaron aramid filament yarn.
Opportunities and challenges		+ Silver-coated polymer thread: unsuitable as electrically conductive control tearing thread, since elongations at break cannot be reproduced or the parallel position of the untwisted filaments results in only individual filaments tearing in case of failure and the applied tension remaining constant − Metallic fine wire: very sensitive to breakage, otherwise excessively high elongation at break compared to load-bearing agent − Cellulose fiber with carbon filling: lower, moisture-dependent conductivity than silver-plated polyamide yarns or fine wires. − Optically conductive control thread: buckling sensitivity, critical mechanical behavior.
MATERIAL PROPERTIES	At 20 °C, maximum percentage shrinkage >30%	
ENERGY SUPPLY		Electric current
RESOLUTION	0–10 s	
SENSITIVITY		
MEASUREMENT RANGE	At 20 °C, shrinkage stress = 0.351–1.755 kg/m^2 (30–150 mg/den)	
TRL	9	9

1 | SENSOR TYPE	Mechanical
2 | MEASURAND	Visual assessment
3 | CONSTRUCTION PRINCIPLE	Thread
4 | GEOMETRY	Linear
5 | MATERIAL	Polypropylene, polyethylene terephthalate

a) Procedural principle

Sensor with optical signal output in the event of critical wear or damage to the outer sheath of a load-bearing rope or tape. The sensor thread has a core-sheath structure, the signal-colored core being sheathe with thermoplastic staple fiber. This has the color of the load-bearing textile and is integrated into its outer shell in such a way that it is exposed to abrasion during use. [4]

b) Schematic sketch

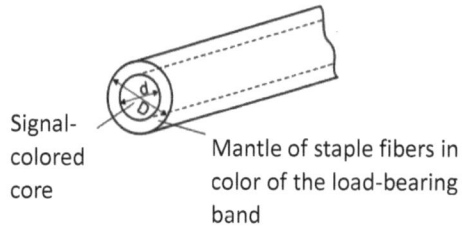

Signal-colored core

Mantle of staple fibers in color of the load-bearing band

c) Known/possible field of application

Structural health monitoring of ropes.

d) Possible sensor variants

Variation of the ratio of core to shell diameter.

Core yarn made of PET, sheath yarn made of PP; core thread not signal-colored, but made of fluorescent material for UV detection.

Variation of core and sheath strength.

Core yarn made of PP, sheath yarn made of PET.

e) Opportunities and challenges

+ The use of a fluorescing signal thread in the thread core enables an automated visual inspection of the wear condition by means of camera technology, even on soiled or very colorful load-bearing textiles
+ With increasing sheath fineness, there is a significant increase in bearable double chafing

I MATERIAL PROPERTIES	
II ENERGY SUPPLY	None
III RESOLUTION	
IV SENSITIVITY	
V MEASUREMENT RANGE	
VI TRL	6–8

	PROTECH	
	MOISTURE- AND CHEMICAL-SENSITIVE SENSOR THREAD	**RAPID-SHRINK FIBER**
SENSOR TYPE	Mechanical, chemical	Chemical
MEASURAND	Visual assessment	Visual assessment
CONSTRUCTION PRINCIPLE	n/a	Warp knit, weave
GEOMETRY	Linear	Linear, planar
MATERIAL	n/a	Elastomer (polyutherane, rubber)
Procedural principle	Permanent identification of harmful environmental influences through the use of threads which change their shape, color or volume while absorbing liquids. The core yarn must be UV-resistant and clearly distinguished in color from the load-bearing tape. For the sheath fibers of the yarn, a material must be selected which is changed in shape, color or structure by UV radiation. [4]	A polymer fiber which shrinks rapidly at ordinary temperature and in contact with water, but retains the fiber shape (impact strength), has high absorbency and has performance characteristics such as rubber elasticity. [14]
Schematic sketch	Core yarn (UV sensor) / Coat fiber (wear sensor)	Radius r / Water-absorbent shrinkable yarn / Rapidly shrinking fiber / Pitch d / Twist angle
Known/possible field of application	A friction-spun sensor thread represents a combination of an abrasion sensor and a UV sensor.	Disposable diapers; fastening tapes; cloths as covers for dampening units in offset printers; cords or cylinders for plant cultivation; cords and nets for the food industry; bank reinforcements.
Possible sensor variants	Decrease in abrasion resistance with increasing exposure to UV radiation.	A water-absorbing, shrinkable yarn produced by blending or blending spinning the rapidly shrinking fiber and a fiber that shrinks slower than said fiber upon absorption of water. A water-absorbing shrinkable material which consists of a water-absorbing shrinkable fibrous web and a water-absorbing shrinkable yarn that absorbs water at a higher rate and to a greater extent than the fibrous web, with the water-absorbent shrinkable yarn containing the rapidly shrinking fiber.
Opportunities and challenges		
MATERIAL PROPERTIES		At 20 °C, maximum percentage shrinkage >30%
ENERGY SUPPLY	None	
RESOLUTION		0–10 s
MATERIAL		
MATERIAL PROPERTIES		At 20 °C, shrinkage stress = 0.351–1.755 kg/m^2 (30–150 mg/den)
TRL	9	9

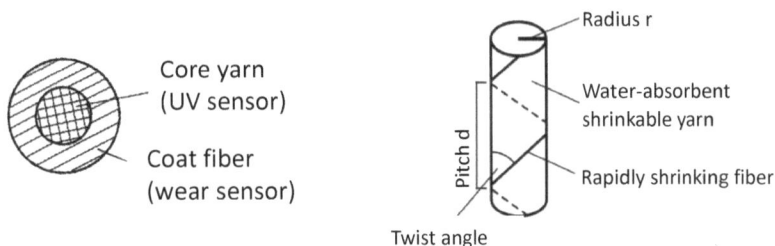

	PROTECH	
	TEMPERATURE CONTROLLED RADIATION TRANSMISSION	**STRAIN/PRESSURE SENSOR**
1 | SENSOR TYPE	Mechanical	Mechanical
2 | MEASURAND	Electric current	Electric current
3 | CONSTRUCTION PRINCIPLE	Fleece	Weft knit, warp knit
4 | GEOMETRY	Flat, fiber diameter: 0.01 to 10 mm	Planar
5 | MATERIAL	Thermotropic polymer blends	Stainless steel
a) Procedural principle	A polymer-based material having temperature-controlled radiation transmission which is present within core/sheath fibers in a core. A transparent shell surrounds the core of thermotropic polymer mixture, which becomes turbid beyond the so-called lower critical demixing temperature (LCST) due to a changing radiation emission. This turbidity effect occurs due to a structural change in the polymer system, in which the components with different refractive indices separate due to temperature change. A variation of the relative contents of the individual comonomers causes turbidity at different temperatures. [17]	Spacer weft-knit made of electrically conductive stainless-steel-fiber yarns for detecting the position of the contact and the size of the contacting surfac when the specific electrical resistance of the electrically conductive conductor paths changes as result of elongation or pressure. [21]
b) Schematic sketch		
c) Known/possible field of application	Temperature-dependent control of radiation transmission on buildings (awnings, roller blinds, venetian blinds), technical equipment, in the clothing industry and for decorative purposes.	Flat pressure load in buildings.
d) Possible sensor variants	Incorporation of a non-thermotropic but mechanically highly resilient material into the polymer core.	Spacer warp-knit can also be used instead of space weft-knit.
e) Opportunities and challenges	+ Advantage of core-shell structure when using aids with low compatibility with thermotropic core material − Expensive production − Bonding of polymers only possible at high application temperatures − Limited possibility of reversible structural change − Low mechanical load capacity	− Spacer warp-knit has a hysteretic force behavior and is therefore less suitable as a pressure senso
I MATERIAL PROPERTIES	Relative proportion of comonmers between 0.1 and 50 mol%	
II ENERGY SUPPLY	Electromagnetically	Electric current
III RESOLUTION		
IV SENSITIVITY		
V MEASUREMENT RANGE		
VI TRL	9	6–8

	TEXTILE NETTLE CELL	TEXTILES WITH SPECIAL FUNCTIONS
SENSOR TYPE	Thermal	Chemical
MEASURAND	Temperature	Visual assessment, electric current
CONSTRUCTION PRINCIPLE	Wire, integrated in support fabric	
GEOMETRY	Planar	
MATERIAL	Shape-memory metal	Hollow polymers modified with moisture-sensitive gels
Procedural principle	Implementation of the textile sensor in the initial fabric by which it autarkically warns the wearer of excessive heat stress on the outside of the garment due to irritation on the inside of the fabric. Heat collectors (metal plates) pass heat onto a heat insulator (time delay element), which delivers a defined amount of heat to a rolled, blunt needle made of shape-memory metal (nitinol). With sufficient heat, the needle stretches through the undergarment and irritates the skin of the wearer. [26]	Garment comprising sensory and/or actuatorically modified polymers which, in the event of a health and/or environmental hazard, change their color, geometric shape or other physical, biological or chemical properties to protect the wearer in a defined manner. [27]
Schematic sketch	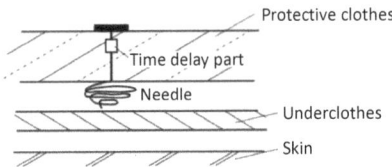	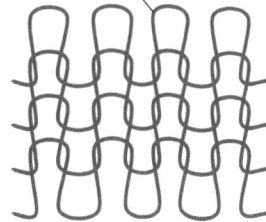
Known/possible field of application	Personnel potentially exposed to high temperatures.	Monitoring of danger conditions.
Possible sensor variants	Simple configuration of sensor sensitivity via material selection.	Sensor element can be formed from: temperature sensors, pressure sensors, humidity sensors, pH sensors, radiation sensor.
Opportunities and challenges	+ Cost-effective and unproblematic made-to-measure clothing + Self-sufficient and redundant system + No susceptible cabling + Fast location and size estimation of the heat source + No warning signals need to be monitored continuously + Having few layers of clothing prevents greater heat stress	
MATERIAL PROPERTIES		
ENERGY SUPPLY	Heat	
RESOLUTION		
MATERIAL		
MATERIAL PROPERTIES		
TRL	6–8	6–8

Schematic sketch (left):
Protective clothes
Time delay part
Needle
Underclothes
Skin

Schematic sketch (right):
Basic thread in right-left-binding

PAGE 180	INDICATIVE COLOR SENSOR	CLOTHING INDICATOR FOR UV RADIATION AND OZONE
1 | SENSOR TYPE	Chemical	Chemical
2 | MEASURAND	Visual assessment	Visual assessment
3 | CONSTRUCTION PRINCIPLE	Printed fabric	
4 | GEOMETRY	Punctiform and areal	
5 | MATERIAL	Fluorescent agent	
a) Procedural principle	A light-sensor layer, temperature sensor layer and fluorescent layer with applied writing, pattern or three-dimensional form, which change their shape and aesthetic impression when externally influenced. [28]	Small-scale application of a variety of fashionable design forms whose color change is accompanied by influencing factors from the environment and a least semi-quantitatively correlates with the hazar potential. The measurement is carried out either b the iodine method, acetone decomposition, oxalic acid decomposition or an IG dosimeter.
b) Schematic sketch	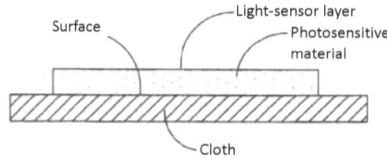	
c) Known/possible field of application	Light-sensor layer for detection of UV radiation. Temperature sensor layer for temperature determination. Fluorescent layer for generating fluorinating light.	Sensor application to swimwear, leisurewear and workwear for outdoor activities for the detection o UV radiation.
d) Possible sensor variants		Determination of the intensity of UV radiation by measurement using iodine method, acetone decay oxalic acid decomposition or IG dosimeter.
e) Opportunities and challenges	+ Optically appealing design of signal bodies	+ Good resistance of the textile carrier to the haza potential, sensitisation technology and reaction causing color change − Doubts as to whether the concentration and intensity of the hazard potential is sufficient to initiate the chemical reaction on the textile
I MATERIAL PROPERTIES		
II ENERGY SUPPLY	Light and heat	Light
III RESOLUTION		
IV SENSITIVITY		
V MEASUREMENT RANGE		
VI TRL	6–8	<6

Schematic sketch labels: Surface, Light-sensor layer, Photosensitive material, Cloth

PROTECH

	ABC PROTECTIVE CLOTHING	PHOSPOHR TEMPERATURE SENSOR
SENSOR TYPE	Mechanical, chemical, thermal	Thermal
MEASURAND	Electric current	Temperature
CONSTRUCTION PRINCIPLE	Weft knit, warp knit, weave, scrim, fleece and composite fabrics	Weave
GEOMETRY	Punctiform, linear or planar	Planar
MATERIAL		MPD-I, PPD-T, PBI
Procedural principle	Protective clothing against biological and chemical toxins and pollutants that warns of exposure to hazards by using at least one sensor and changing its electrical properties. [29]	Multilayer garment whose outer sheath is equipped with a temperature sensor. This ensures rapid expansion of the textile under the effect of heat to ensure a heat-insulating intermediate layer to protect the body. [30]
Schematic sketch		Eu-dopted phosphor Glass fiber
Known/possible field of application	PPE	Protective clothing (coat, jacket, trousers) for firefighters or industrial applications under high heat exposure.
Possible sensor variants	Sensor element can be formed from: temperature sensors, pressure sensors, humidity sensors, pH sensors, radiation sensor.	A 25% increase in time until second-degree burn on the skin occurs compared to conventional protective clothing.
Opportunities and challenges	+ Indication of the end of the recommended wearing period + No time limit for the wearing period	
MATERIAL PROPERTIES		Riskiness from 1.5 to 4
ENERGY SUPPLY	Electric current	Heat
RESOLUTION		0–3 s
SENSITIVITY		
MEASUREMENT RANGE		
TRL	6–8	9

PROTECH

	PHYSIOLOGICAL SENSOR 1	PHYSIOLOGICAL SENSOR 2
1 | SENSOR TYPE	Chemical, mechanical	Mechanical
2 | MEASURAND	Electric current	Electromagnetic light spectrum
3 | CONSTRUCTION PRINCIPLE	Conductive yarn	Fiber, fiber braid, diffraction grating
4 | GEOMETRY	Planar	Optical fiber with outer diameter of 125 µm, grating diameter of 6–9 µm, sensor diameter of 150–250 µ
5 | MATERIAL	Silver/gold-coated nylon	Polymers, glass, electrically conductive metals
a) Procedural principle	Recording of physiological states via sensors, with transmission via electrically conductive conductor paths in clothing and processing in measuring equipment. [31]	A patient-monitoring system comprising a plurality of diffraction gratings arranged along an optical fiber. Each optical fiber and grating is configured to change either the effective refractive index or the grating periodicity of the corresponding grating at its location along the fiber in response to at least one desired external stimulus. [41]
b) Schematic sketch		
c) Known/possible field of application	Sportswear and medical clothing for monitoring bodily functions. Multimedia clothing for adapting media enjoyment to physiological conditions.	Nursing of newborns.
d) Possible sensor variants	Sportswear/medical clothing. [32], [33], [34], [35], [36] Textile electrode in spacer warp-knit. [37], [38], [39] Multifunctional apparel system. [40]	Reduced number of required connection options.
e) Opportunities and challenges	+ Advantageous contact behavior due to pressure-elastic behavior when using monofilaments + Acceptance by the wearer due to attractive appearance + Comfortable to wear due to the flexibility of the garment	+ Hygiene + Skin sensitivity + Wear acceptance
I MATERIAL PROPERTIES	Electrical resistance <5 Ω/cm; diameter of monofilaments >100 µm	
II ENERGY SUPPLY	Electric current	Electric current
III RESOLUTION		
IV SENSITIVITY		
V MEASUREMENT RANGE		
VI TRL	9	9

Labels in schematic sketch: Bragg grid; Protective layer; Optical fiber core

	PHYSIOLOGICAL SENSOR 3	PHYSIOLOGICAL SENSOR 4
SENSOR TYPE	Thermal	Mechanical, chemical, thermal
MEASURAND	Electromagnetic light spectrum	Electric current
CONSTRUCTION PRINCIPLE	Weft knit, warp knit, weave	Elastic weave or fleece equipped with electrically conductive fibers
GEOMETRY	Sheath diameter: 0.125 mm; core diameter: 0.09 mm	Linear, planar
MATERIAL		Elastomers filled with conductive particles or electrically conductive metals
Procedural principle	Integration of a fiber-optic temperature-sensing element into a fabric. The temperature sensing element is an optical fiber containing one or more fiber-Bragg-grating sensors. Light is introduced into the optical single-mode fiber and directed to a grating interface adjacent to the wearer. A reflux signal is received by a reflection mode or a transmission mode, with the reflux signal having a wavelength shift indicative of temperature by the Bragg resonance effect. [42]	Garment with belts running transversely to the longitudinal axis of the wearer, which can be stretched in the longitudinal direction and in which strain gauges are incorporated, which allow physiological functions to be determined by changing the electrical conductivity. [43]
Schematic sketch	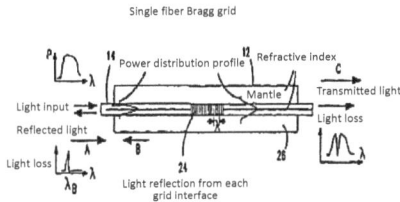	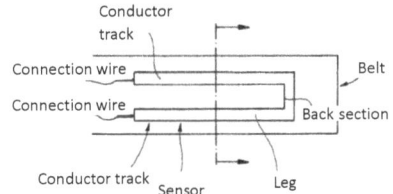
Known/possible field of application		Clothing for monitoring heart activity and recording skin resistance, perspiration and body temperature.
Possible sensor variants	Processing of the thread in a weft knit, warp knit or weave.	The carrier material of the electrically conductive threads is knitted fabric made of cotton with elastane content or viscose, or synthetic or microfiber. Conductive particles in the elastor of the strain sensor can be carbon particles or hydrogels.
Opportunities and challenges		+ The garment should be resistant to perspiration and at the same time resistant to washing + Increase of sensor sensitivity through path-shaped guidance of the strain sensor, since the transverse elongation is low compared to a longitudinal elongation − An insulating layer should prevent moisture from influencing the measuring signal of the extensometers − The elastomer should be more extensible than the substrate on which the sensor is placed so that the extensibility of the sensor does not limit that of the garment
MATERIAL PROPERTIES		Specific sensor resistance: 5–30,000 Ωcm
ENERGY SUPPLY	Electric current	Electric current
RESOLUTION		
SENSITIVITY		
MEASUREMENT RANGE		
TRL	6–8	9

	PHYSIOLOGICAL SENSOR 5	PHYSIOLOGICAL STRAIN SENSOR
1 | SENSOR TYPE	Mechanical	Mechanical
2 | MEASURAND	Electric current	Electric current
3 | CONSTRUCTION PRINCIPLE	Weft knit provided with electrically conductive threads	Elastic fabric provided with electrically conductive filaments
4 | GEOMETRY	Punctiform, linear or planar	Linear, planar
5 | MATERIAL		
a) Procedural principle	Sensor consisting of strain gauges, piezoelectric elements, length gauges or pressure sensors, all of which change their electrical properties under mechanical deformation. [44]	Electrically conductive thread for determining the state of respiration and movement, which changes its electrical properties under tensile or compressive load, above all its electrical resistance and inductance. [46]
b) Schematic sketch		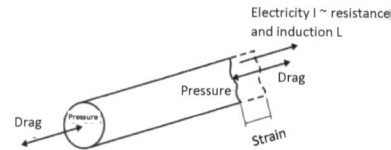
c) Known/possible field of application	Garment for determining a posture or movement of the body.	Clothing for monitoring respiratory and physical activity of newborns, children, adults and even nor human mammals.
d) Possible sensor variants	Sensor element can be formed from strain gauges.	Clothing for monitoring respiratory and physical activity of newborns, children, adults and even nor human mammals.
e) Opportunities and challenges	+ Piezoelectric elements + Magnetic, capacitive or optical length gauges. Pressure sensors + High wearing comfort due to unobtrusive integration of the sensor elements into the garment	+ Moisture resistant, i.e., washable + High accuracy of measurement + No slipping of the sensors due to precise positioning in the garment + Increased wearing comfort due to the not too tight fit of the garment
I MATERIAL PROPERTIES		
II ENERGY SUPPLY	Electric current	Electric current
III RESOLUTION		
IV SENSITIVITY		
V MEASUREMENT RANGE		
VI TRL	9	9

	PROTECH	
GE 185	KNITTED BREATHING SENSOR	THREE-DIMENSIONAL SPACER WARP-KNIT
SENSOR TYPE	Mechanical	Mechanical
MEASURAND	Electric current	Electric current
CONSTRUCTION PRINCIPLE		Warp knit
GEOMETRY		Planar
MATERIAL	Electrically conductive polyester with stainless-steel content	
Procedural principle	Measurement of respiratory movement by changing the electrical resistance when weft knits made of conductive polyester fiber yarns with stainless-steel content are stretched. [48]	Three-dimensional spacer warp-knit with integrated ultrasonic sensors for monitoring body movement. [50]
Schematic sketch	Electricity I ~ resistance R and induction L / Drag / Pressure / Drag / Pressure / Strain	Electrically conductive fiber / Polymer fiber / Poorly conductive polymer shell / **Longitudinal section**
Known/possible field of application		
Possible sensor variants	Right/left weft-knit with conductive stripes. Right/left lining weft-knit in which the conductive yarn no longer forms any stitches, but is merely integrated with handles in the non-conductive basic knit Right/right weft-knits where the electrical resistance is less dependent on elongation	Flexible elastane material guarantees flexibility and wearing comfort.
Opportunities and challenges		
MATERIAL PROPERTIES		
ENERGY SUPPLY	Electric current	Electric current
RESOLUTION		
SENSITIVITY		
MEASUREMENT RANGE		
TRL	6–8	6–8

	PROTECH	
	SHAPE-MEMORY SENSOR	**SAFETY CLOTHING**
1 | SENSOR TYPE	Thermal	Mechanical
2 | MEASURAND	Electric current, temperature	Electric current
3 | CONSTRUCTION PRINCIPLE	Wire, integrated in support fabric	Weft knit provided with electrically conductive threads
4 | GEOMETRY	Planar	Linear, planar
5 | MATERIAL	Metallic alloys, polymers	Electrically conductive metals
a) Procedural principle	By heating to a certain temperature via an electric current, the fabric takes on a desired shape with integrated conductive wires. When the electric current is deactivated, the material returns to its original shape. [51]	Garment for locating stab wounds or gunshot wounds to the human body by using sensor units arranged in electrical conductor tracks, the operating principle of which is the piezoelectric effect. [52]
b) Schematic sketch	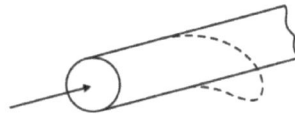 Electricity ~ temperature	
c) Known/possible field of application	Protective suit for pilots of fighter planes who are exposed to high forces on the body due to high acceleration values.	Protective vests for the police and military.
d) Possible sensor variants	Disposable shape-memory effect: by only one phase transition in the metallic alloy, the material can only reach its original state. Two-way shape-memory effect: two different original material states can be achieved by varying the temperature into a high and a low temperature.	Variation of the conductor arrangement, preferably in wave or curve form, as these ensure an elastic arrangement. The construction of many smaller circuits enables a more precise location of the interruption of the conductor path and thus of the injury to humans. Polymer tracks can be printed, embroidered or woven directly onto the fabric.
e) Opportunities and challenges	+ Fast reaction time + Functional maintenance even with minor damage + The total weight and installation space of the device are less than those corresponding to the state of the art − Permanent irreversible plastic deformation of up to 0.1%	+ Detection of impacts, pressure waves and detonations using piezoelectric elements
I MATERIAL PROPERTIES		
II ENERGY SUPPLY	Electric current	Electric current
III RESOLUTION	0.2–1 s	
IV SENSITIVITY		
V MEASUREMENT RANGE	<100 °C	
VI TRL	9	9

	PROTECH	
	INTELLIGENT SKIN ARCHITECTURE	**PRESSURE SENSOR**
SENSOR TYPE	Mechanical	Chemical
MEASURAND		Electromagnetic light spectrum
CONSTRUCTION PRINCIPLE	Weave	
GEOMETRY	Planar	Linear
MATERIAL		
Procedural principle	Optical fibers woven into a carrier material which serve as sensors for optical information transmission. [53]	Realization of a pressure sensor with the help of two aligned optical waveguides (one fixed, the other movable).
Schematic sketch		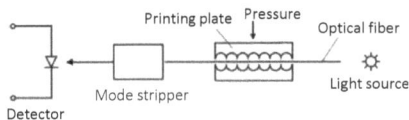
Known/possible field of application	Acquisition of data; image processing; communication.	
Possible sensor variants	Supporting weaving of the optical fibers into channels. Arrangement of the optical fibers in a grid-like mat consisting of fibers of any carrier material. Woven structure comprising a first group of warp-direction yarns and a second group of weft-direction yarns with optical fibers arranged between selected pairs of the first group. Optoelectronic packaging structure with two sections, in each of which the abovementioned woven structure is placed.	Pressure measurement using the "microbending-effect", in which small deviations of the optical fiber axis from a straight line cause mechanical stresses in the core and cladding, which in turn cause light to be decoupled.
Opportunities and challenges	+ Low construction volume; low weight + High tensile strength; high elasticity; high resistance to weathering; high resistance to chemicals; high tear strength; high dimensional stability; high wear resistance − Sensitivity to deflection, leading to a deterioration in transmittance	
MATERIAL PROPERTIES		
ENERGY SUPPLY	None	Light
RESOLUTION		
SENSITIVITY		0–20 bar
MEASUREMENT RANGE		
TRL	6–8	6–8

	SPORTTECH	
	INDICATIVE COLOR SENSOR	**PHYSIOLOGICAL SENSOR 1**
1 | SENSOR TYPE	Chemical	Chemical, mechanical
2 | MEASURAND	Visual assessment	Electric current
3 | CONSTRUCTION PRINCIPLE	Printed fabric	Conductive yarn
4 | GEOMETRY	Punctiform and areal	Planar
5 | MATERIAL	Fluorescent agent	Silver/gold-coated nylon
a) Procedural principle	A light-sensor layer, temperature sensor layer and fluorescent layer with applied writing, pattern or three-dimensional form, which change their shape and aesthetic impression when externally influenced. [28]	Recording of physiological states via sensors, transmission via electrically conductive conductor paths in clothing and processing in measuring equipment. [31]
b) Schematic sketch		
c) Known/possible field of application	Light-sensor layer for the detection of UV radiation. Temperature sensor layer for temperature determination. Fluorescent layer for generating fluorinating light.	Sportswear and medical clothing for monitoring bodily functions. Multimedia clothing for adapting media enjoyment to physiological conditions.
d) Possible sensor variants		Sportswear/medical clothing. [32], [33], [34], [35], [36] Textile electrode in spacer warp-knit. [37], [38], [39] Multifunctional apparel system. [40]
e) Opportunities and challenges	+ Optically appealing design of signal bodies	+ Advantageous contact behavior due to pressure-elastic behavior when using monofilaments + Acceptance by the wearer due to attractive appearance + Comfortable to wear due to the flexibility of the garment
I MATERIAL PROPERTIES		Electrical resistance <5 Ω/cm; diameter of monofilaments >100 µm
II ENERGY SUPPLY	Light and heat	Electric current
III RESOLUTION		
IV SENSITIVITY		
V MEASUREMENT RANGE		
VI TRL	6–8	9

Schematic sketch labels: Surface, Light-sensor layer, Photosensitive material, Cloth

	SPORTTECH	
	PHYSIOLOGICAL SENSOR 4	**PHYSIOLOGICAL SENSOR 5**
SENSOR TYPE	Mechanical, chemical, thermal	Mechanical
MEASURAND	Electric current	Electric current
CONSTRUCTION PRINCIPLE	Elastic weave or fleece provided with electrically conductive fibers	Weft knit provided with electrically conductive threads
GEOMETRY	Linear, planar	Punctiform, linear or planar
MATERIAL	Elastomers filled with conductive particles or electrically conductive metals	
Procedural principle	Garment with belts running transversely to the longitudinal axis of the wearer, which can be stretched in the longitudinal direction and in which strain gauges are incorporated, which allow physiological functions to be determined by changing the electrical conductivity. [43]	Sensor consisting of strain gauges, piezoelectric elements, length gauges or pressure sensors, all of which change their electrical properties under mechanical deformation. [44]
Schematic sketch		
Known/possible field of application	Clothing for monitoring heart activity and recording skin resistance, perspiration and body temperature.	Garment for determining a posture or movement of the body.
Possible sensor variants	The carrier material of the electrically conductive threads is knitted fabric made of cotton with elastane content or viscose, or synthetic or microfiber. Conductive particles in the elastor of the strain sensor can be carbon particles or hydrogels.	Sensor element can be formed from strain gauges.
Opportunities and challenges	+ The garment should be resistant to perspiration and washing + Increase of sensor sensitivity through path-shaped guidance of the strain sensor, since the transverse elongation is low compared to a longitudinal elongation − An insulating layer should prevent moisture from influencing the measuring signal of the extensometers − The elastomer should be more extensible than the substrate on which the sensor is placed so that the extensibility of the sensor does not limit that of the garment	+ Piezoelectric elements + Magnetic, capacitive or optical length gauges; pressure sensors + High wearing comfort due to unobtrusive integration of the sensor elements into the garment
MATERIAL PROPERTIES	Specific sensor resistance: 5–30,000 Ωcm	
ENERGY SUPPLY	Electric current	Electric current
RESOLUTION		
MATERIAL		
MATERIAL PROPERTIES		
TRL	9	9

The schematic sketch labels: Conductor track, Connection wire, Connection wire, Conductor track, Sensor, Belt, Back section, Leg

1 | SENSOR TYPE	Chemical
2 | MEASURAND	Electromagnetic light spectrum
3 | CONSTRUCTION PRINCIPLE	
4 | GEOMETRY	Linear
5 | MATERIAL	

a) Procedural principle

Realization of a pressure sensor with the help of two aligned optical waveguides (one fixed, the other movable).

b) Schematic sketch

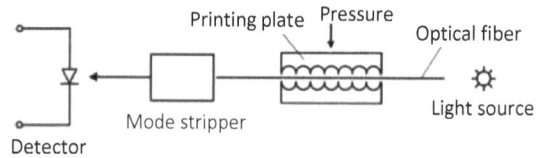

Printing plate Pressure Optical fiber Light source Mode stripper Detector

c) Known/possible field of application

d) Possible sensor variants

Pressure measurement using the "microbending-effect", in which small deviations of the optical fiber axis from a straight line cause mechanical stresses in the core and cladding, which in turn cause light to be decoupled.

e) Opportunities and challenges

I MATERIAL PROPERTIES	
II ENERGY SUPPLY	Light
III RESOLUTION	
IV SENSITIVITY	0–20 bar
V MEASUREMENT RANGE	
VI TRL	6–8

	SPORTTECH	
	THREE-DIMENSIONAL SPACER WARP-KNIT	**SHAPE-MEMORY SENSOR**
SENSOR TYPE	Mechanical	Thermal
MEASURAND	Electric current	Electric current, temperature
CONSTRUCTION PRINCIPLE	Warp knit	Wire, integrated in support fabric
GEOMETRY	Planar	Planar
MATERIAL		Metallic alloys, polymers
Procedural principle	Three-dimensional spacer warp-knit with integrated ultrasonic sensors for monitoring body movement. [50]	By heating to a certain temperature via an electric current, the fabric takes on a desired shape with integrated conductive wires. When the electric current is deactivated, the material returns to its original shape. [51]
Schematic sketch		
Known/possible field of application		Protective suit for pilots of fighter planes who are exposed to high forces on the body due to high acceleration values.
Possible sensor variants	Flexible elastane material guarantees flexibility and wearing comfort.	Disposable shape-memory effect: by only one phase transition in the metallic alloy, the material can only reach its original state. Two-way shape-memory effect: two different original material states can be achieved by varying the temperature into a high and a low temperature.
Opportunities and challenges		+ Fast reaction time + Functional maintenance even with minor damage + The total weight and installation space of the device are less than those corresponding to the state of the art − Permanent irreversible plastic deformation of up to 0.1%
MATERIAL PROPERTIES		
ENERGY SUPPLY	Electric current	Electric current
RESOLUTION		0.2–1 s
SENSITIVITY		
MEASUREMENT RANGE		<100 °C
TRL	6–8	9

1	SENSOR TYPE	Mechanical

2	MEASURAND	Electric current

3	CONSTRUCTION PRINCIPLE	Fiber optic

4	GEOMETRY	Linear conductor, fiber length between 100 and 1000 m

5	MATERIAL	

a) Procedural principle

Ring interferometer which evaluates the phase difference between the opposing light waves, which is dependent on the angular velocity, as a measured variable. Polarized laser light passes between two beam splitters before it is coupled into the two ends of the same fiber coil. In the case of a stationary system, light paths of equal lengths of the circulating modes result in a constructive interference at the output of the second beam splitter, whereas a destructive interference occurs at the output of the first beam splitter. The relativistic Sagnac effect results in a phase difference $\Delta\Phi$ between the light waves rotating in opposite directions, which is proportional to the product of the conversion number m and the enclosed area A. [1]

b) Schematic sketch

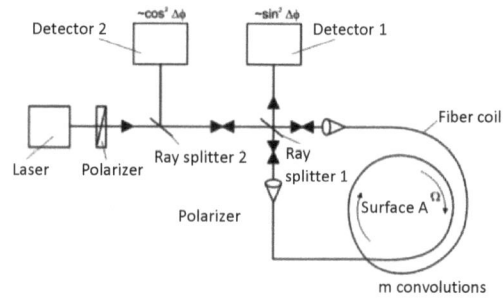

c) Known/possible field of application

Earth rotation measurement.

Navigation tools.

Robot control.

d) Possible sensor variants

Integrated optical resonator: sensitivities up to several 100s of °/h.

e) Opportunities and challenges

+ Miniaturization of the fiber-optic gyroscope through integrated optics
+ Use in areas with short-term stability as well as with required long-term stability possible

I	MATERIAL PROPERTIES	

II	ENERGY SUPPLY	Laser light

III	RESOLUTION	

IV	SENSITIVITY	Up to 3–10 °/h

V	MEASUREMENT RANGE	

VI	TRL	9

Sohler, W. Faseroptische und integriert optische Sensoren. In: *Sensoren in der textilen Meßtechnik*; Eckhard Schollmeyer, E.; Hemmer, A., Eds.; Book Series Fachberichte Messen, Steuern, Regeln, Volume 12; Springer-Verlag: Berlin/Heidelberg, Germany, New York/Toronto, USA, 1985; p. 55ff.

Kunert Fashion GmbH & Co. Strickware mit Feuchtigkeitssensor. DE 10 2009 052 929 A 1, 19 May 2011.

Nageldinger, G. Optoelektronischer Sensor zur Erfassung von anhaftenden Flüssigkeitsanteilen in oder an flüssigkeitsspeichernden Substanzen. DE 41 16 633 A1, 26 November 1992.

Möhring, U.; Scheibner, W.; Gries, T.; Stüve, J. *Erhöhung der Funktionssicherheit von gewebten und geflochtenen lastaufnehmenden Bändern und Seilen für industrielle Anwendungen und Extremsportbereiche durch Integration von Sensoren für die Belastungs- und Verschleißkontrolle*; Schlußbericht für den Zeitraum 01.04.2004–30.06.2006; Textilforschungsinstitut Thüringen- Vogtland e.V. Greiz, Institut für Textiltechnik der RWTH Aachen, Germany.

Oeste, F.D. Haas, R. Detektoren zur Wasseruntersuchung. DE 299 23 893 U1, 23 August 2001.

Oeste, F.D. Haas, R. Detektionsmittel und ihre Anwendungen. DE 199 47 635 A1, 25 January 2001.

Kabelwerk Oberspree GmbH. Faseroptischer Sensor zum Nachweis von gasförmigen oder flüssigen Medien. DE 041 22 619 A1, 14 January 1993.

Vorbach, D.; Taeger, E. Eigenschaften von kohlenstoffgefüllten Cellulosefasern. *Technische Textilien* **1998**, 41, 2, 67–70.

Daniel, E. Einsatz von Lichtwellenleitern in der Textilveredlung. *TEMA Technik und Management* **1990**, 40, 1, 34–36.

Kurt-Schwabe-Institut für Mess- und Sensortechnik e.V. Faseroptischer pH-Sensor. DE 20 2006 011 421 U1, 7 December 2006.

GESO Gesellschaft für Sensorik. Vorrichtung und Verfahren zur Kontrolle und Überwachung der Dämmung von Mantelrohren. DE 198 43 974 C1, 20 April 2000.

Nageldinger, G. Optoelektronischer Sensor zur Erfassung von anhaftenden Flüssigkeitsanteilen in oder an flüssigkeitsspeichernden Substanzen. DE 41 16 633 A1, 26 November 1992.

Commissariat à l'Energie Atomique. Faseroptischer, aktiver, chemischer Sensor und Herstellungsverfahren desselben. DE 00 0069 015 487 T2, 27 July 1995.

Kuraray Co. Schnellschrumpfende Faser, wasserabsorbierendes schrumpfendes Garn und andere, diese Faser enthaltende Gegenstände. DE 00 0003 687 735 T 2, 9 September 1993.

Wunderlich, K. Unsichtbarer Schutz, Alarmtapeten zur Flächenüberwachung Wirtschaftsschutz & Sicherheitstechnik. *Zeitschrift für das Sicherheitswesen in der Wirtschaft* **1999**, 6, 60–61.

Sächsisches Textilforschungsinstitut. Lamelle zur Ertüchtigung und Überwachung von Tragwerken sowie Verfahren zu deren Herstellung und Anwendung. DE 10 2008 052 807 B3, 25 February 2010.

BASF AG. Material mit temperaturgesteuerter Strahlungsemission. DE 198 19 368 A1, 4 November 1999.

Müller, W. Hybridseil für Hub- und Transporteinrichtungen insbesondere für Aufzüge. DE 202 02 989 U1, 18 July 2002.

Hegger, J.; Molter, M.; Hofmann, D.; Basedau, F.; Gutmann, T.; Habel, W. Verbunduntersuchungen an textilbewährten Betonkörpern mittels faseroptischer Mikrodehnungssensoren. *VDI-Berichte* **2001**, 1599, 165–170.

Hanselka, H. Adaptive Faserverbunde wirken sensorisch und aktorisch, Schwingungen und Lärm im Keim erstickt. *Industrieanzeiger* **2000**, 2. Available online: https://industrieanzeiger.industrie.de/allgemein/schwingungen-und-laerm-im-keim-erstickt/ (accessed on 19 March 2019).

Tillmanns, A.; Heimlich, F.; Brücken, A.; Weber, M.O. Abstandsgestricke als Drucksensoren. *Technische Textilien* **2009**, 3, 136.

Adeloka, A.; Richter, A.-N. Membrankonstruktionen der dritten Generation. *Textilforum* **1995**, 3, 34–35.

Gries, T.; Bosowski, P.; Hörr, M.; Mecnika, V.; Jockenhövel, S. Design and manufacture of textile-based sensors. In *Electronic Textiles: Smart Fabrics and Wearable Technology*; Dias, T., Ed.; Woodhead Publishing Series in Textiles 166; Woodhead Publishing: Cambridge, UK, 2015; pp. 75–107.

Zangani, D.; Fuggini, C.; Loriga, G. Electronic textiles for geotechnical and civil engineering. In *Electronic Textiles: Smart Fabrics and Wearable Technology*; Dias, T., Ed.; Woodhead Publishing Series in Textiles 166; Woodhead Publishing: Cambridge, UK, 2015; pp. 275–300.

Lorussi, F.; Carbonaro, N.; Rossi, D. de; Tognetti, A. Strain- and Angular-Sensing Fabrics for Human Motion Analysis in Daily Life. In *Smart Textiles: Fundamentals, Design, and Interaction*; Schneegass, S., Ed.; Springer: Cham, Switzerland, 2017; pp. 49–70.

Institut für Textiltechnik der RWTH Aachen. Textile Nesselzelle zur Frühwarnung vor Überhitzung von Schutzbekleidung. DE 103 08 238 A 1, 2 September 2004.

Ahlers, H., Doz. Dipl.-Ing. habil. Textilien mit Sonderfunktionen. DE 00 0019 619 858 B 4, 29 April 2004.

B&R WIN CO., LTD. Kleidung mit Sensor. DE 20 2008 016 426 U 1, 28 May 2009.

Blücher GmbH. Funktionelles Bekleidungsstück, insbesondere ABC-Bekleidung, mit integrierter Meßeinrichtung. DE 20 2006 001 661 U 1, 15 March 2007.

E.I. du Pont de Nemours and Co. Schutzbekleidung. DE 694 17 757 T 2, 11 November 1999.

Deutsche Telekom AG. Intelligente Kleidung. DE 00 0010 047 533 A 1, 11 April 2002.

32. Boson Technology Co., Ltd. Sportbekleidung mit der Möglichkeit zur Erfassung des physiologischen Zustands. DE 20 2010 001 514 U1, 20 May 2010.

33. Koninklijke Philips Electronics N.V. Kleidungsstück mit eingearbeiteten medizinischen Sensoren. DE 602 24 717 T2, 8 January 2009.

34. Cairos technologies AG. Kleidungsstück zum Überwachen physiologischer Eigenschaften. DE 10 2008 051 536 A1, 15 April 2010.

35. Quantum Applied Science and Research Inc. Kleidungsstück mit eingepassten physiologischen Sensoren. DE 10 2005 026 897 A1, 29 December 2005.

36. Mega Elektroniikka Oy; Suunto Oy. Sensoranordnung zum Messen von Signalen auf der Hautoberfläche und Verfahren zur Herstellung der Sensoranordnung. DE 11 2004 001 921 T5, 28 August 2008.

37. Militz, D. Textilelektrode. DE 20 2006 007 226 U1, 11 October 2007.

38. Elastic Textile Europe GmbH. Textile Flächenelektrode sowie Kleidungsstück mit einer solchen. DE 10 2009 028 314 A1, 10 February 2011.

39. Daimler AG. Textilelektrode. DE 10 2008 049 112 A1, 7 May 2009.

40. Sächsisches Textilforschungsinstitut e.V. Multifunktionales sensorintegriertes Bekleidungssystem. DE 20 2005 021 140 U1, 13 September 2007.

41. General Electric Co.Vorrichtungen und System zur faseroptischen Multiparamter-Patientengesundheits Überwachung. DE 10 2009 003 37 A 1, 30 July 2009.

42. General Electric Co. Temperatur messendes Gewebe. DE 10 2008 013 052 A 1, 18 September 2008.

43. Deutsche Institute für Textil- und Faserforschung (DITF). Kleidungsstück mit integrierter Sensorik. DE 10 2004 030 261 A 1, 19 January 200

44. Infineon Technologies AG. Kleidungsstück mit integrierten elektronischen Komponenten. DE 00 0010 350 869 A 1, 9 June 2005.

45. OFA Bamberg GmbH. Faden zur Ermittlung der Zugspannung, insbesondere in einem medizinischen Gestrick oder Gewirk. DE 10 2008 00. 122 A 1, 9 July 2009.

46. Fraunhofer-Gesellschaft zur Förderung der angewandet Forschung e.V. Kleidungsstück zur Erfassung einer Atmungsbewegung. DE 10 200 053 843 A 1, 20 May 2009.

47. Alpha-Fit GmbH. Drucksensor. DE 10 2005 055 842 A 1, 24 May 2007.

48. Ehrmann, A.; Heimlich, F.; Brücken, A.; Weber, M.O.; Haug, R. Gestrickter Atemsensor. *Melliand Textilberichte* **1999**, 6,10, 234–235.

49. Lilienfeld-Toal, H. von. Druckempfindlicher Strumpf. DE 00 0010 314 211 A 1, 6 November 2003.

50. Schindlbeck, K.; Heide, M. Frühwarngerät gegen Rückenschmerzen, Sensortechnik im Abstandsgewirke hält den Rücken unter Kontrolle. *Kettenwirk-Praxis* **2002**, 3, 17.

51. Celsius Aerotech A.B; Linköping, S.E. Vorrichtung zur Ausübung eines externen Druckes auf einen Körper. DE 695 13 153 T 2, 17 August 200

52. Dynacc GmbH. Bekleidungsstück für einen menschlichen Körper. DE 10 2009 046 861 A 1, 26 May 2011.

53. Page Automated Telecommunications Systems. Intelligente Hautarchitektur von gewobenem, faseroptischen Flachbandarchitekturen un Verfahren zu ihrer Verpackung. DE 692 24 444 T2, 7 January 1999.

54. Park, S.-H.; Lee, H.B.; Yeon, S.M.; Park, J.; Lee, N.K. Flexible and Stretchable Piezoelectric Sensor with Thickness Tunable Configuration of Electrospun Nanofiber Mat and Elastomeric Substrates. *ACS applied materials & interfaces* **2016**, 8, 37, 24773–24781.

55. Miao, M. Carbon nanotube yarns for electronic textiles. In *Electronic Textiles: Smart Fabrics and Wearable Technology*; Dias, T., Ed.; Woodhe Publishing Series in Textiles 166; Woodhead Publishing: Cambridge, UK, 2015; pp. 55–72.

56. Zhou, B.; Lukowicz, P. Textile Pressure Force Mapping. In *Smart Textiles: Fundamentals, Design, and Interaction*; Schneegass, S., Ed.; Spring Cham, Switzerland, 2017; pp. 31–47.

57. Hoechst AG. Polyimidwellenleiter als optische Sensoren. DE 39 26 604 A1, 14 February 1991.

58. Höer, J.; Wetter, O.; Peschke, C.; Schumpelick, V.; Weck, M. Einsatz miniaturisierter Sensoren zur Kontrolle von Naht- und Fadenspannung der Chirurgie. In *Wissenschaftliche Gesellschaft für Produktionstechnik: 3. Symposium: Neue Technologien für die Medizin, Forschung – Praxis – Innovation, Aachen, 10 May–11 May 2001*; Weck, M., Ed.; Shaker: Aachen, Germany, 2001; 323–340.

59. Reuber, C. Lichtwellenleiter Entwickler suchen Glasfaser- Anwendungen Einsatzchancen zwischen Rechner und Maschine und Sensoren. *VDI- Nachrichten* **1984**, 38, 47, 18.

60. Suh, M. Wearable sensors for athletes. In *Electronic Textiles: Smart Fabrics and Wearable Technology*; Dias, T., Ed.; Woodhead Publishing Series in Textiles 166; Woodhead Publishing: Cambridge, UK, 2015; pp. 257–273.

61. Carl Freudenberg KG. Kombiniertes Sensor- und Heizelement. DE 103 58 793 A1, 4 August 2005.

62. W.E.T. Automotive Systems AG. Mehrschichtiges übernähtes System. DE 10 2004 022 373 B4, 16 March 2006.

MDPI

St. Alban-Anlage 66

4052 Basel

Switzerland

Tel. +41 61 683 77 34

Fax +41 61 302 89 18

www.mdpi.com

MDPI Books Editorial Office

E-mail: books@mdpi.com

www.mdpi.com/books

MDPI

www.ingramcontent.com/pod-product-compliance
Lightning Source LLC
Chambersburg PA
CBHW051559190326
41458CB00029B/6481